MEDIA DYNAMICS IN SOUTH ASIA

SERIES EDITORS

Adrian Athique, Vibodh Parthasarathi, and S.V. Srinivas

The centrality of mediation to social, economic, and political processes in South Asia has become increasingly evident in recent years. The media has become a substantial economic sector, with considerable strategic and symbolic importance for career decisions and for the fortunes of South Asia in the global economy. The intense mediation of the political arena has seen a wide range of media formats taking on the role of actors in the day-to-day operation of the democratic process and in the conduct of international relations. The use of media technologies for enabling the interlocking fields of education, employment, and consumption makes their functions and potentials a necessary concern for the social sciences and humanities. Beyond their utility, the expressive content of the media requires a deep engagement with cultural reproduction across the region.

Media Dynamics in South Asia curates an interdisciplinary approach that addresses media studies as a field of interlocking interests in politics, economics, culture, technology, gender, and education. Our imperative for expanding the breadth of media studies in this way serves the larger task of uncovering the various relationships, transactions, and interactions that characterize social change as a dynamic process. At the same time, our broad sociological approach to different aspects of the media requires us to account in depth for the conditions, concerns, and challenges specific to the region. It is no longer sufficient to maintain a paradigm for media analysis conceived elsewhere and transplanted, often uncomfortably, to various locations in the subcontinent.

Rather, the presence of a large, sophisticated, and fast-moving media environment in South Asia promises sufficient depth to support original and innovative research approaches and the production of a future teaching curriculum grounded in the regional experience. The series seeks to play a critical role in establishing the necessary resources for supporting the growth of media studies in South Asia.

Adrian Athique is associate professor of cultural studies at the Institute for Advanced Studies in the Humanities, the University of Queensland, Brisbane, Australia.

Vibodh Parthasarathi is a founding faculty and associate professor at the Centre for Culture, Media and Governance, Jamia Millia Islamia, New Delhi, India.

S.V. Srinivas is professor at the School of Liberal Studies, Azim Premji University, Bengaluru, India.

EDITORIAL BOARD

MEDIA DYNAMICS IN SOUTH ASIA

The Politics of Digital India

Between Local Compulsions and Transnational Pressures

Pradip Ninan Thomas

OXFORD
UNIVERSITY PRESS

OXFORD
UNIVERSITY PRESS

Oxford University Press is a department of the University of Oxford.
It furthers the University's objective of excellence in research, scholarship,
and education by publishing worldwide. Oxford is a registered trademark of
Oxford University Press in the UK and in certain other countries.

Published in India by
Oxford University Press
2/11 Ground Floor, Ansari Road, Daryaganj, New Delhi 110 002, India

ISBN-13 (print edition): 978-0-19-949462-0
ISBN-10 (print edition): 0-19-949462-2

ISBN-13 (eBook): 978-0-19-909785-2
ISBN-10 (eBook): 0-19-909785-2

Typeset in Adobe Jenson Pro 10.5/13 by
by The Graphics Solution, New Delhi 110 092
Printed in India by Nutech Print Services India

To friends and friendships:
Zaharom Nain (Rom), Ramesh Ramanathan (Zamby), and Philip Lee

Contents

Acknowledgements

There are numerous critical scholars in the political economy tradition who have influenced my work and provided me with insights into the politics of the digital, such as Peter Golding, Graham Murdock, and Adrian Athique, who played no small part in the shaping of this manuscript, along with Vibodh Parthasarathi, Dan Schiller, Nic Carah, Tom O'Regan, and Zaharom Nain, among many others.

May your tribe increase!

I would also like to thank the editorial team at Oxford University Press for its help.

Abbreviations

A2K	Access to Knowledge (movement)
AATP	Agricultural Advanced Technology Park
ABD	accumulation by dispossession
AEBPR	Advanced eBook Processor
AFTI	Alliance for Fair Trade with India
AIDS	acquired immune deficiency syndrome
ALAC	At-Large Advisory Committee
AP	Andhra Pradesh
ASSOCHAM	Associated Chambers of Commerce and Industry of India
ATM	automated teller machine
BJP	Bharatiya Janata Party
BKS	Bharatiya Kisan Sangh
BOO	build–own–operate
BOSS	Bharat Operating System Solutions
BOT	build–operate–transfer
BPO	business process outsourcing
BRICS	Brazil, Russia, India, China, and South Africa
C-DAC	Centre for Development of Advanced Computing
ccTLD	country code top-level domain
CCTNS	Crime and Criminal Tracking Network and Systems
CCTV	closed-circuit television
CEO	chief executive officer
CERT	Computer Emergency Response Team

CIDR	Central Identities Data Repository
CII	computer-implemented inventions
CIRP	Committee for Internet Related Policies
CIS	The Centre for Internet & Society
CMS	Central Monitoring System
CRIS	Communication Rights in the Information Society
CRUSH	Criminal Reduction Utilizing Statistical History
CSC	Computer Science Corporation
CSTD	Commission on Science and Technology for Development
DAISY	Digital Accessible Information System
DeitY	Department of Electronics and Information Technology
DIY	do it yourself
DMCA	Digital Millennium Copyright Act
DNA	deoxyribonucleic acid
DNS	Domain Name System
DRM	Digital Rights Management
DSCI	Data Security Council of India
DTTI	Defense Technology and Trade Initiative
ECIL	Electronics Corporation of India
EDS	Electronic Data Systems
EICTIA	European Information and Communication Technology Industry Association
EU	European Union
4G	fourth generation
FDI	foreign direct investment
FICCI	Federation of Indian Chambers of Commerce and Industry
FOSS	free and open source software
GAC	Government Advisory Committee
GATT	General Agreement on Trade and Tariffs
GDP	gross domestic product
GEAC	Genetic Engineering Appraisal Committee
GIFT City	Gujarat International Finance Tec-City
GIS	geographical information systems
GM	genetically modified
GoI	Government of India

GPS	geographical positioning systems
GSLV	Geosynchronous Satellite Launch Vehicle
GSTN	Goods and Services Tax Network
gTLD	generic top-level domain
HIV	human immunodeficiency virus
IAB	Internet Architecture Board
IADC	International Ad Hoc Committee
IANA	Internet Assigned Numbers Authority
IBSA	India–Brazil–South Africa
ICANN	Internet Corporation for Assigned Names and Numbers
ICT	information and communications technology
ID	identity [related to identification schemes]
IDN	internationalized domain name
IETF	Internet Engineering Task Force
IFSEC	International Fire and Security Exhibition and Conference
IFWP	International Forum on the White Paper
IG	Internet governance
IGF	Internet Governance Forum
IIGF	India Internet Governance Forum
IIM	Indian Institute of Management
IIT	Indian Institute of Technology
IMF	International Monetary Fund
IP	intellectual property
IPC	Intellectual Property Committee
IPv6	Internet Protocol version 6
ISPs	Internet service providers
ISPAI	Internet Service Providers Association of India
ISRO	Indian Space Research Organisation
IT	information technology
ITRs	International Telecommunications Regulations
ITU	International Telecommunication Union
KIA	Knowledge Initiative on Agriculture
KRRS	Karnataka Rajya Raitha Sangha
MEA	Ministry of External Affairs
MMP	Mission Mode Project
MNC	multinational corporation

MoU	memorandum of understanding
MSH	multistakeholderism
NAM	National Association for Manufacturers
NASA	National Aeronautics and Space Association
NASSCOM	National Association of Software and Services Companies
NATGRID	National Intelligence Grid
NATO	North Atlantic Treaty Organization
NCTC	National Counter-Terrorism Centre
NETRA	Network Traffic Analysis Initiative
NGO	non-governmental organization
NIS	National Identity Scheme (UK)
NITI Aayog	National Institute for Transforming India Aayog
NIXI	National Internet Exchange of India
NPR	National Population Register
NRCFOSS	National Resource Centre for Free/Open Source Software
NSA	National Security Agency
NTIA	National Telecommunications and Information Administration
NWEO	New World Economic Order
NWICO	New World Information and Communication Order
NWIO	New World Information Order
OADA	Open Ag Data Alliance
OECD	Organisation for Economic Co-operation and Development
OSDD	Open Source Drug Discovery
OSSI	Open Source Seed Initiative
PDS	public distribution system
PIL	public interest litigation
PM	prime minister
PNAS	*Proceedings of the National Academy of Sciences of the United States of America*
POTA	Prevention of Terrorism Act
PPP	public–private partnership
PPV	Protection of Plant Varieties
PPVFRA	Protection of Plant Varieties and Farmers' Rights Act
PSLV	Polar Satellite Launch Vehicle

PTLB	Perry4Law Techno Legal Base
PUCL	People's Union for Civil Liberties
PwC	PricewaterhouseCoopers
R&D	research and development
RSS	Rashtriya Swayamsevak Sangh
RTI	Right to Information [Act]
SCoF	Standing Committee on Finance
SEZs	special economic zones
SFLC	Software Freedom Law Centre
SITE	Satellite Instructional Television Experiment
SJM	Swadeshi Jagran Manch
SMS	short message service
SPACE	Society for Promotion of Alternative Computing and Employment
STPI	Software Technology Parks of India
3G	third generation
TCP/IP	Transmission Control Protocol/Internet Protocol
TCS	Tata Consultancy Services/TeleCommunication Systems, Inc.
TLD	top-level domain
TPMs	technological protection measures
TRAI	Telecom Regulatory Authority of India
TRIPS	Trade-Related Aspects of Intellectual Property Rights
TV	television
UAV	unmanned aerial vehicle
UID	Unique Identity (project)
UIDAI	Unique Identification Authority of India
UK	United Kingdom
ULBs	urban local bodies
UN	United Nations
UNESCO	United Nations Educational, Scientific and Cultural Organization
UPOV	International Union for the Protection of New Varieties of Plants
URLs	Uniform Resource Locators
USA	United States of America
USAID	United States Agency for International Development

USDA	United States Department of Agriculture
USPTO	United States Patent and Trademark Office
USSR	Union of Soviet Socialist Republics
USTR	Office of the United States Trade Representative
VHP	Vishva Hindu Parishad
VIPs	visually impaired persons
W3C	World Wide Web Consortium
WBU	World Blind Union
WCIT	World Conference on International Telecommunication
WIPO	World Intellectual Property Organization
WSIS	World Summit on the Information Society
WTO	World Trade Organization

Introduction

Observations on the Politics of Digital India

This book explores the politics and geopolitics of digital India. While the politics of digital India can be explored from a variety of vantage points, I seek to highlight a range of critical, current 'information' issues that are being shaped by both internal and external political and geopolitical imperatives. In considering the consequences of the shaping of the knowledge economy in India, I attempt to address current issues such as surveillance, software patenting, algorithmic power, the political economy of seed, biotechnology and the digital, and Internet governance, along with the ways in which these issues are currently being negotiated, contested, and/or resolved both nationally and with international institutions such as the Internet Corporation for Assigned Names and Numbers (ICANN) and countries such as the United States of America (USA). Why are these specific areas related to the digital highlighted in this text? It is because these are the areas that are key to India's strategic interests and reflect its ambition to become a knowledge superpower. I consider seed and precision agriculture precisely because the digital footprint has progressively claimed every productive sector within its ambit. The politics and geopolitics of the digital, as dealt with in this book, have two dimensions:

1. Geopolitics, and especially the relationship between India and the USA, and the ways in which policies and practices related to the digital are shaped in India as a consequence of this relationship; and
2. The internal/domestic consequences of digital policy and practice that are contested and shaped as much by local contingencies as by global fiat.

I highlight the fact that the particularities of the Indian context simply have not been accounted for in policymaking related to digital India—a context in which the agricultural economy continues to account for more than half of the workforce. In 2017, less than 4 million out of the 450 million workforce were employed in information technology (IT) and IT-enabled services, and very modest numbers were involved in offering services to these industries as drivers, caterers, guards, domestic workers, and so on.[1]

The political dimensions of digital India clearly indicate the fact that the role of the State is key to understanding the nature of both public and private platforms in India. There are, however, formidable interested parties and trade partners, such as the USA, who leverage their power to shape India's digital future. While the Indian government has made numerous attempts to shape the specific contours of the digital knowledge economy in India, most recently through the 'Digital India' initiative that was launched in 2015, the US government has attempted to leverage its power to shape cross-sectoral informational futures in India through the means of bilateral trade and Section 301 provisions. As such, the shaping of India's knowledge futures is deeply contested from both within and without, and this book explores some of these contestations. Using insights from both Karl Polanyi and David Harvey to understand the nature of the neo-liberal state, its contradictions and contestations, I argue that the reform and redistributive agenda related to the digital emerges from within the State and from outside it, specifically communities, such as farmers and social movements. To date, this larger context is scarcely acknowledged in official initiatives such as Digital India. Rather, the heady language of Digital India tends to elide these divides in favour of a discourse that views the digital as a benign force that is autonomous and disembedded from the social, but which has the intrinsic power to

---

[1] For a critique of official employment figures in this sector, see Barnes, T. 2015. 'The IT Industry, Employment and Informality in India: Challenging the Conventional Narrative', *The Economic and Labour Relations Review* 26(1): 82–9. See also Preetha, S. and A.J. Solanki. 2017. 'India: A New Challenge for India's Technology Sector: Trade Union Registered in Bangalore', *Mondaq*, 17 November, available at http://www.mondaq.com/india/x/647374/employee+rights+labour+relations/A+new+challenge+for+Indias+technology+sector+Trade+union+registered+in+Bangalore; accessed on 23 May 2018.

transform the country into a modern, digital nation and its millions of people into 'digital natives'.

The Geopolitics of Digital India

The geopolitics of India has traditionally been studied against the background of the British Empire and in the context of the Cold War. India has always maintained its strategic interest in manufacturing indigenous technologies for deployment in both development and defence. However, India's eschewal of socialism and embracing of capitalist futures, globalization, consumerism, its traditions of democracy, and its double-digit growth rates in the IT sector have led to positive reassessments of the country as a global player. Undoubtedly, the single most important development that has contributed to the country's newly found international status is the development of its service economy, namely, business, banking and communication, retail and wholesale trade and transport, and a variety of household-related service industries. The services sector contributed 53.8 per cent to the total gross domestic product (GDP) in 2016–17, out of which 7.7 per cent has been attributed to the IT–business process outsourcing (BPO) sectors, a fall from the steady growth experienced from 2008–9, with a peak of 9.5 per cent in 2014–15. In 2012–13, the total revenue earned through software exports was $75.8 billion, while the total revenue earned by the IT industry in 2017 was $116 billion. Indeed, the IT industry proclaimed that it employed 3 million, while 9 million were indirectly employed.[2] It is evident that while employment rates in the IT and ancillary industries remain insignificant when compared to agriculture (close to 230 million) and the manufacturing industry (92 million), the share of the GDP that the IT sector brings in is significant.[3]

The establishment of a single, integrated, and all-encompassing knowledge economy in India that is harmoniously integrated with global expectations, norms, and rules is a fraught project as it is also shaped by a variety of 'local' compulsions. India's ambition of 'strategic autonomy' in

[2] Statista. 2017. 'Direct and Indirect Employment of the IT-BPM Industry in India from Fiscal Year 2009 to 2017 (in Millions)', available at http://www.statista.com/statistics/320729/india-it-industry-direct-indirect-employment/.

[3] Statista. 'Direct and Indirect Employment of the IT-BPM Industry in India from Fiscal Year 2009 to 2017 (in millions)'.

its geopolitical dealings makes it an especially interesting case given that there are many examples of the Indian government rather unambiguously supporting local interests, such as the 'compulsory licensing' of pharma products, which are deemed to be in the public interest but are antithetical to trade rules and counter to the expectations of international pharmaceutical companies. This book explores these global–local compulsions and the many ways in which the 'local' is adapting the global to its own interests. Within this narrative, the role played by the Indian state in negotiating the politics of information is rarely consistent and is based on reconciling the interests of multinational corporations (MNCs) such as Microsoft and Facebook, specific interest groups such as the IT sector (through enabling software patents), political partners such as the present government's[4] negotiations with its family of Hindu organizations that belong to the Sangh Parivar, and its citizens. The controversy related to Facebook's stillborn Free Basics initiative is interesting precisely because the current prime minister (PM) was a vocal supporter of this initiative, although he did not envisage the nature of internal opposition within the government, notably the Telecom Regulatory Authority of India (TRAI) and nationwide advocacy by the civil society.

This episode highlights the fact that the knowledge economy in India is being shaped by multiple compulsions and, as such, there are no guarantees that the end result will be a replica of policy blueprints or reflect the visions of assorted MNCs who have a stake in India's digital economy. Arguably, the Government of India (GoI) has to work within a fractured consensus on the exact shaping of digital India and the need to balance the interests of various stakeholders, notably industry and its citizenry. The Indian government is, on the one hand, committed to economic liberalization and to opening India to the global market exemplified by the Digital India initiative, while it is also, at the very same time, unable to control the many local digital economies, the so-called 'jugaad' economies, that have sprung up and are outside the 'formal' digital economy. In other words, the paradoxes and contradictions of the Indian state as it negotiates the knowledge economy are many. In its avatar as a security state, it is involved in the control of its citizenry through the expansion of surveillance via its Unique Identity (UID) or Aadhaar project, although it

---

[4] The term 'present government' in this volume refers to the NDA government (2014–19).

has, at a local level, also been supportive of open source software solutions in governance and education that have led to the empowerment of citizens in select states in India inclusive of Karnataka and Kerala. The GoI is beholden to the IT industry and uses the privilege of the state to acquire land for the establishment of special economic zones (SEZs). At the same time, it occasionally supports policies and initiatives that are in the public interest, such as the amendment to the copyright law that has enabled greater access to digital material for the visually impaired and global advocacy at WIPO for stronger access-based policies for the visually impaired.

The State has also had to deal with the fact that the main political challenge to e-commerce has come from within the Hindutva family, which includes the current ruling party, the Bharatiya Janata Party (BJP). The Swadeshi Jagran Manch (SJM), the economic advocacy unit of Rashtriya Swayamsevak Sangh (RSS)—an all-India cadre-based Hindu nationalist organization—is vociferously opposed to the major e-commerce vendors currently operating in India, including the foreign-funded Indian company Flipkart and the impact of such vendors on the offline retail market.[5] A brainchild of the RSS chief, Balasahib Deoras, the SJM was formed to counter the liberalization policies adopted by the Narasimha Rao government in 1991. The SJM, embarrassingly enough for the government, has launched a campaign against Paytm, a digital wallet company financially supported by Alibaba—a reflection of its growing anti-China stance following the stand-off between India and China in Doklam in 2017.[6] The SJM has also been involved in a campaign against the government regulator Genetic Engineering Appraisal Committee's (GEAC) support for genetically modified (GM) mustard and the pro-pharma policies adopted by National Institute for Transforming India, or NITI Aayog, the successor to the Planning Commission, and the Public

[5] See Mathew, L. 2015. 'An RSS Reminder: Ban e-Tail like Amazon, Flipkart', *The Indian Express*, 18 January, available at http://indianexpress.com/article/india/india-others/meanwhile-an-rss-reminder-ban-e-tail-like-amazon-flipkart/; accessed on 11 May 2018.

[6] PTI. 2017. 'RSS-Affiliated Swadeshi Jagran Manch up in Arms against Paytm', *Livemint*, 5 November, available at http://www.livemint.com/Politics/vKTtPPKve86RL3OiEYuWJM/RSSaffiliated-Swadeshi-Jagran-Manch-up-in-arms-against-Payt.html; accessed on 22 May 2018.

Health Foundation of India, a non-governmental organization (NGO) supported by funds from the Bill and Melinda Gates Foundation.[7]

The multifaceted response by the State reflects Lloyd Rudolph and Susanne Rudolph's descriptor of the Indian state as 'polymorphous' and a 'creature of manifold forms and orientations' in their book, *The Pursuit of Lakshmi: The Political Economy of the Indian State*.[8] While their own understanding of the Indian state as a self-determining entity was shaped by the independent 'dirigiste', developmentalist state in the post-Independence period, their book did not foreshadow the Indian state's formal eschewal of socialism and its embrace of the neo-liberal model of growth that began in the 1990s and has accelerated ever since. In other words, their analysis deals with the developmentalist state before it was recalibrated by the logic of the market in the early 1990s. The recalibrated Indian state has downgraded its ideational and material support for welfarism and *Garibi Hatao* ('End Poverty', one of the Congress Party's electoral slogans during PM Indira Gandhi's reign) and has since focused on establishing policies and infrastructure that enable capitalist growth. In Priya Chacko's words:

> The erosion of the developmentalist state project from the 1970s laid the path for a deeper shift in the national social order from the 1990s with the recasting of statehood within a geo-economic social, wherein India's future was thought to be best secured through policies of economic openness, growth and competitiveness, rather than through geo-political social forms of central planning and endogenous economic development. This shift in India's state project has given rise to new forms of global and regional engagement that have served to further processes of state transformation in India.[9]

[7] Mahajan, A.S. 2017. 'The Battle Within: The Modi Government's Reform Agenda Continues to Face Opposition from the RSS', *Business Today*, 12 February, available at https://www.businesstoday.in/magazine/features/the-modi-governments-reform-agenda-continues-to-face-opposition-from-the-rss/story/244680.html; accessed on 11 March 2019.

[8] Rudolph, L.I. and S.H. Rudolph. 1987. *In Pursuit of Lakshmi: The Political Economy of the Indian State*. Chicago and London: University of Chicago Press, p. 400.

[9] Chacko, P. 2014. 'The New Geo-Economics of a "Rising" India: State Transformation and the Recasting of Foreign Policy', *Journal of Contemporary Asia* 45(2): 341.

This erosion of the developmentalist state is also reflected in the State's relentless cuts in public expenditures, agricultural subsidies, rural development, investments in agriculture, in elementary education, and health, with the latter steadily undergoing privatization. India's post-Independence commitment to capitalism, which has been accentuated in the period of economic liberalization, still remains a deeply uneven process. In D'Costa's words: 'capitalism in India must be seen for what it is, a market system that is evolving in an uneven way, geographically, sectorally and socially.'[10]

India's embrace of the market and the knowledge economy has been accompanied by a shift in its international relations as it looks towards the West, the USA in particular—a change from its alignment with the Union of Soviet Socialist Republics (USSR) during the Cold War period. It is undoubtedly India's relationship with the USA that is key to understanding its digital economy and geopolitics. Over the previous two decades, there has been a perceived convergence of interests between the USA and India, especially in areas of security, strengthening of democracy, and embrace of policies supportive of neo-liberal growth in India.[11] Not only is the USA a key trading partner and the major destination for software exports from India, it is also home to some of the largest information and media companies that now have a substantive presence in India and which employ vast numbers of Indian employees in their industries, both within India and elsewhere. This relationship has recently become strained as a result of the USA's increasing adoption of protectionist policies and restrictions to the issuance of H1B visas, which were issued predominantly to those of Indian origin working in the IT sector. The consequent slowdown in software exports and the general malaise in the IT sector in India have led to redundancies in key IT companies in India, including Infosys, Cognizant, and Tech Mahindra, with projected job losses in excess of 200,000 between 2017 and 2020.[12]

[10] D'Costa, A.P. 2005. *The Long March to Capitalism: Embourgeoisement, Internationalisation and Industrial Transformation in India*. Basingstoke and New York: Palgrave Macmillan, p. 7.

[11] See Chitalker, P. and D.M. Malone. 2011. 'Democracy, Politics and India's Foreign Policy', *Canadian Foreign Policy Journal* 17(1): 75–91.

[12] Srivastava, S. 2017. 'India Tech Sector Downsizes Heavily as Trump's H-1B Policies Create Uncertainty', *CNBC*, 23 May, available at https://www.

This relationship is also reflected in the shifts in the Indian state's 'moral' leadership: from a country that supported pro-Independence struggles in Palestine and Tibet to one that is actively involved as an economic partner of Israel and China. The relationship with Israel is especially interesting, given India's launch of an advanced Israeli spy satellite in 2008, Israeli involvement in providing surveillance systems to help with India's security and monitoring operations in Kashmir, and the ongoing cooperation in other areas such as agricultural development.[13] There have been major bilateral initiatives, including

> the India–Israel Industrial R&D and Technological Innovation Fund (I4F), the India–Israel CEO forum, the India–Israel Innovation Bridge, an online platform to encourage and facilitate collaboration between Israeli and Indian start-ups; the Indo-Israeli Agriculture Project; MoUs (memorandums of understanding) between the Indian Space Research Organization (ISRO) and the Israel Space Agency (ISA); and MoUs on India–Israel water cooperation.[14]

The Indian state's present orientation is also reflected in its continuing partnerships with the Anglo Commonwealth nations, especially in building nuclear capabilities based on cooperation with the United Kingdom (UK), Canada, and Australia.[15]

The State and Its Digital Entanglements

I have, in other writings, described the State in India as an ambivalent entity, torn between the imperatives of shaping India's future along neo-liberal

---

cnbc.com/2017/05/23/indian-tech-sector-downsizes-heavily-as-trumps-h-1b-policy-creates-uncertainty.html; accessed on 12 May 2018.

[13] Saint-Mezard, I. 2010. 'India and Israel: An Unlikely Alliance', *Le Monde Diplomatique*, November, available at https://mondediplo.com/2010/11/11indiaisrael; accessed on 22 May 2018.

[14] Mishra, D. 2018. 'Future Prospects of the India–Israel Defense Cooperation', *The Jerusalem Post*, 1 January, available at http://www.jpost.com/Opinion/Future-prospects-of-the-India-Israel-defense-cooperation-522586; accessed on 22 May 2018.

[15] Davis, A.E. 2014. 'The Identity Politics of India–US Nuclear Engagement: Problematising India as Part of the Anglosphere', *Journal of the Indian Ocean Region* 10(1): 81–96.

pathways on the one hand, and the redistribution of resources to large sections of the Indian population on the other.[16] One of the characteristic features of the neo-liberal State in India is that it has become thoroughly 'marketized', perhaps best illustrated by the fact that the Digital India project is based on public–private partnership (PPP) in which private firms are closely involved in shaping India's digital future. Furthermore, the involvement of key representatives of India's major business houses in party politics and their election to the Rajya Sabha—for example, Rahul Bajaj (BJP/Shiv Sena and National Congress Party), Anil Ambani (Samajwadi Party), and Rajkumar Dhoot (Shiv Sena)—and lobbying for their own corporate interests, points to the systems of patronage that continue to frame State–market relationships.[17] The ability of the State to regulate the market is, in the context of this close relationship, difficult to say the least, precisely because of analogous interests. The appointment of IT Secretary R.S. Sharma, a key advocate of Aadhaar and other schemes linked to Digital India, as chairperson of TRAI in 2015 has been reported as a political appointment, although he too could not prevent the setback of Facebook's 'Free Basics' and the cause of 'net neutrality' that was upheld by TRAI in 2016.[18]

The previous Congress-led government had also attempted to use the offices of TRAI to push through a 2G spectrum auction that was itself a major financial scam that had direct consequences for the political fortunes of that government.[19] Both examples point to the fact that the power of the State to support its own extra-legal interests, and that of the market, has remained constant irrespective of the regime in power. At a macro level, this pro-market orientation is reflected in and across the various tiers

[16] Thomas, P.N. 2014. 'The Ambivalent State and the Media in India: Between Elite Domination and the Public Interest', *International Journal of Communication* 8: 466–82.

[17] Srivastava, J. 2013. 'Theorising State–Market Axis in a Globalizing India', Paper presented at the ECPR General Conference, Bordeaux, 4–7 September, p. 12, available at https://ecpr.eu/Filestore/PaperProposal/40d3e6fe-2f58-4459-ac56-7d0871d62217.pdf; accessed on 11 July 2018.

[18] Bhatia, R. 2016. 'The Inside Story of Facebook's Biggest Setback', *The Guardian*, 12 May, available at https://www.theguardian.com/technology/2016/may/12/facebook-free-basics-india-zuckerberg; accessed on 10 July 2018.

[19] Bhandari, B. 2012. *Spectrum Grab: Inside Story of the 2G Grab*, New Delhi: BS Books.

of the developmental state. The replacement of the Planning Commission with NITI Aayog in 2015, at least on paper, was defended as a necessary exercise that would result in the replacement of a command economy with an economy based on cooperate federalism that would strengthen India's strategic economic interests within a competitive global economy. While there certainly is evidence of competitive federalism making a difference—for example, the success of the IT sectors in Karnataka and Telangana—the lack of federal financial support for Amaravati, the de facto capital of/smart city project in Andhra Pradesh (AP), does suggest that there are limits to cooperative federalism, one of the key principles underlying the vision and mission of NITI Aayog. Arguably, such examples of an impasse between the central government and states can lead to the expansion of competitive federalism.[20] Regional disparities are exemplified by the fact that Uber, IKEA, DreamWorks, Google, Amazon, and Apple have invested in their biggest campuses outside of the USA in Telangana but not in any of the other states.

While the central government is, at least in principle, committed to an all India approach to the 'digital', it is clear that the digital footprint in India in terms of infrastructure, foreign direct investment (FDI) inflows, social capital, and so on, is massively uneven—disparities that are a direct consequence of varying state investments in education, infrastructure, and connectivity. The much-lauded growth in Internet penetration rates, the distribution of mobile phones, and digital access belies a real consideration of the fact that the epicentres of India's knowledge economy are state and city specific. Seven states accounted for 70 per cent of FDI flows during April–December 2017 (Delhi, Haryana, Maharashtra, Karnataka, Tamil Nadu, Gujarat, and AP) totalling $35.9 billion, out of which $6.14 billion went to telecommunications, $5.16 billion to computer software and hardware, and $4.6 billion to services,[21] thereby confirming trends noted previously by Lawrence Saez.[22]

[20] Saez, L. 1999. 'India's Economic Liberalization, Interjurisdictional Competition and Development', *Contemporary South Asia* 8(3): 323–45.

[21] Kota, H.B. 2018. 'Tackle Regional Disparity in FDI', *The Pioneer*, 21 March, available at https://www.dailypioneer.com/columnists/oped/tackle-regional-disparity-in-fdi.html; accessed on 10 July 2018.

[22] See Saez.'India's Economic Liberalization, Interjurisdictional Competition and Development'.

Theorizing the Indian State, both in its own right and in terms of its compact with the digital, is complex. If India can no longer be characterized as a unitary, developmental state, is it worth our effort to come to an understanding of the Indian state as a macro-entity? Is the State in India an 'imaginary institution', as argued by Sudipta Kaviraj?[23] Or is it the case that in order to understand the contemporary State in India, we simply have to study the State as an entity that is imbricated in the lives of Indians at different levels, registers, moments, and via different strategies and processes, in spite of efforts by the State to create India-wide centralized systems such as the Aadhaar database? In other words, should we begin from an understanding of what digital India means to ordinary Indians living in specific parts of the country, experiencing variegated levels of access and engagements with the digital? As Sailen Routray has argued:

> Apart from the theoretical benefits, it is also methodologically helpful to do this, as this move makes multiple ethnographic sites available to us for studying the state, its effects and the way governmental programmes and policies get shaped on the ground. This is not merely a methodological or theoretical move unrelated to the real-life morphings of the state. What has been termed as the neo-liberal state seems to be characterised by a certain amount of institutional dispersal and contradictions.[24]

Another important specificity of the central state in its embrace of the digital is its continuing investments in building indigenous capacities in the digital, be it via investments at the Centre for Development of Advanced Computing (C-DAC) for the development of the free software/Linux-based Bharat Operating System Solutions (BOSS) that has become the basis for the development of local language-specific software and multilingual computing, or for satellite-based remote sensing, signal distribution, and monitoring technologies by the Indian Space Research Organisation (ISRO). Whether this public interest dimension of the State will remain in the context of its relentless embrace of the market

---

[23] Kaviraj, S. 2010. *The Imaginary Institution of India*. Hyderabad: Permanent Black/Orient Blackswan.

[24] Routray, S. 2015. 'The Post-Development Impasse and the State in India', *Third World Quarterly* 36(10): 1914.

is anybody's guess; however, it does seem that technologies that enhance security and the strategic interests of the State will continue to be supported. The fact that ISRO's commercial arm, Antrix Corporation, has become a revenue earner in its own right, based on its transponder leasing operations and satellite launching business (Rs 275 crore [2,750 million] in 2016–17 from the Polar Satellite Launch Vehicle [PSLV], Geosynchronous Satellite Launch Vehicle [GSLV], and GSLV Mk-III), is a reflection of another avatar of the Indian State as a market.[25] Arguably, therefore, any attempt to understand this complexity with the aid of established theoretical frameworks must contend with the kaleidoscopic terrain of activities, actionings, pressures, and interests from a multiplicity of stakeholders that are involved in shaping growth and development through both legal and extra-legal means. Unions, movements, business and corporate regimes, lobby groups, and to a lesser extent, civil society groups are involved in pressuring the State to work at their behest, in the pursuit of, as the case may be, accumulation and/or redistribution.

An excellent example of the study of the Indian state, from the bottom up as it were, is Akhil Gupta's study of bureaucracy, structural violence, and poverty, based on the ethnographic fieldwork of two welfare projects in Mandi, Muzaffarnagar district, Uttar Pradesh.[26] He explores the paradoxes of how a 'caring' state embraces, via its bureaucracy, 'uncaring' as a 'constitutive modality' and the structural violence that is its key outcome.[27] Gupta sees the State in terms of 'congeries of institutions and agencies and agendas at different levels that are not necessarily well connected with each other';[28] in other words, a disaggregated notion of the State that has differential, uneven impacts on people divided by poverty and their ability to exercise power and control over key resources. This rather more fluid, conjunctural notion of the State experienced by people in India in their everyday lives can be contrasted with the social imaginary of the

[25] *The Indian Express*. 2018. 'India's Share in Global Satellite Launch Services Goes Up', 29 January, available at http://www.newindianexpress.com/business/2018/jan/29/indias-share-in-global-satellite-launch-services-goes-up-1765068.html; accessed on 10 July 2018.

[26] Gupta, A. 2012. *Red Tape: Bureaucracy, Structural Violence and Poverty in India*. Hyderabad: Orient BlackSwan.

[27] Gupta. *Red Tape*, p. 23.

[28] Gupta. *Red Tape*, p. 55.

all-powerful, unitary Indian state. I have tried to account for at least some of the tensions between these two understandings of the relationship between the State, the digital, and its users in the various chapters of this book, in particular the 'formal' and 'informal' nature of digital access and use. I have attempted to capture some of the ambivalence in all these relationships, at the many levels at which the State, the federated state, the market, and, in this case, the digital compact and its registers operate.

A Theoretical Framework for Understanding the Context of Digital India: David Harvey and Karl Polanyi

The ambivalent and uneven circumstances of digital India prompt my engagement with a theoretical framework drawn from the works of David Harvey and Karl Polanyi, who, in different ways, have dealt with the State's acts of omission and commission in the context of the growing influence of the market. Their contributions towards understanding the nature of the liberal state and the neo-liberal economy, as well as tendencies from both within and without to democratize access to resources, can be usefully applied to an understanding of the digital economy in the context of uneven development. In the case of the latter, Athique, Parthasarathi, and Srinivas have recently invoked Polanyi in an intuitionalist account of the Indian Media Economy, although their opening volumes do not address the political dimensions of the State as it pertains to digitization.[29] Given my own long-standing interest in trying to understand the contested nature of digital India, I have chosen to apply Polanyi's 'double movement' and Harvey's 'accumulation by dispossession' (ABD) to understand the specifics of this contestation. Admittedly, these concepts cannot account for all the complexities that exist between the State, the market, and the digital, although they certainly help clarify some of the broader trends that qualify this relationship. I use these conceptual pegs here to illumine some trends related to the larger politics of digital access and use.

[29] Adrian A., V. Parthasarthi, S. V. Srinivas (eds). 2018a. *The Indian Media Economy: Volume 1: Industrial Dynamics and Cultural Adaptation*. New Delhi: Oxford University Press; and Athique, A., V. Parthasarathi, and S.V. Srinivas (eds). 2018b. *The Indian Media Economy: Volume 2: Market Dynamics and Social Transformations*. New Delhi: Oxford University Press.

My aim is not to apply a theoretical framework that can account for all the complexities of India's engagements with the digital, but instead to engage with a number of entry points that can form the basis for understanding the nature of the digital in India against both the presence and absence of the role of the State and some of its external, 'geopolitical', and internal challenges. In doing so, I will place equal emphasis upon David Harvey's extensive arguments on politics of dispossession and resistance in the context of the liberal state and its welfare agenda. From this perspective, the key to understanding the making of digital India is the role played by the neo-liberal market economy. Post-Independence, India's socialist turn in the 1970s and the subsequent return to the embrace of the market economy that began in the late 1980s (and has accelerated ever since) has been accompanied by varying understandings of the role of the State in relation to the market. The attempt at ending poverty (garibi hatao) in which the state played a pre-eminent role in 'redistribution' has been replaced by a commitment to the knowledge economy in which the private sector is involved in shaping economic growth, while the state is involved in ensuring that the digital mode of production becomes the pre-eminent mode for all transactions between the market and society and the foundation for production and reproduction. In the context of uneven development, the State's compact with the digital economy has met with resistance—resistance that is manifested both within the state and in the informal sector, leading to social movements that have attempted to curb the excessive influence of the dominant market through opening up opportunities for multiple, alternative engagements with the digital.

Both Karl Polanyi (1886–1964) and David Harvey (1935–), in some of their key works, place the central emphasis upon the market. In the case of Polanyi, he debunks the myth of the self-regulating, independent market through a historical account, arguing for its recalibration through state-based welfarist interventions.[30] Addressing the period of high globalization at the turn of the century, Harvey examines the unregulated neo-liberal market based on financialization and privatization, and the various forms of resistance to this logic by social movements.[31] While

---

[30] Polanyi, K. 1944. *The Great Transformation: The Political and Economic Origins of Our Time.* Boston: Beacon Press.

[31] Harvey, D. 2007. 'Neoliberalism and Creative Destruction', *The Annals of the American Academy* 610: 22–44.

Polanyi and Harvey have reflected on the eras that they have lived through, their observations do throw light on the nature of both the welfare economy and the neo-liberal economy in India, as well as attempts from below to democratize access to the digital, its formats, places, and spaces. While David Harvey's concept of ABD can be used to make sense of the larger political economy of the digital in India, Karl Polanyi's assessments of market fundamentalism and counter-movements directed towards curbing its excesses can also be drawn upon to make sense of the contested nature of informational capitalism. Polanyi, in *The Great Transformation*, provided a strong critique of market fetishism and market fundamentalism, of a market that is often described as a stand-alone entity that is disembedded from the 'social' and social relationships.[32] Polanyi has also highlighted what he refers to as a 'double movement': capitalism's excesses that are met with counter-movements aimed at bringing back some equilibrium in the market-wage labour relationship and the place of the 'social'.

While there is a need to recognize differences in the intellectual foci of Polanyi and Harvey, the correspondences are strong, in particular, the shared understandings of causes and consequences of the 'double movement'. This is best illustrated by passages from Harvey: 'The creation of the neoliberal system has entailed much destruction, not only of prior institutional frameworks and powers ... but also of divisions of labour, social relations, welfare provisions, technological mixes, ways of life, attachments to the land, habits of the heart, ways of thought and the like.'[33] Harvey notes that resistance to neo-liberalism has been extraordinarily varied: 'The variety of such struggles was and is simply stunning. It is hard to even imagine connections between them. They were and are all part of a volatile mix of protest movements.'[34] Together, Harvey and Polanyi contribute to a broad conceptual understanding of the contradictory impulses of both state and market, which can be used to make sense of ground realities in contemporary India. The chapters in this book situate their case studies in the broader context of the State's embrace of neo-liberalism, the many realities of dispossession, as well as the counter-movements oriented towards re-embedding the informational economy

---

[32] Polanyi. *The Great Transformation*.
[33] Harvey. 'Neoliberalism and Creative Destruction', pp. 23, 39.
[34] Harvey. 'Neoliberalism and Creative Destruction', pp. 22–44.

within the social. In doing so, it is important that we take cognizance of the fact that not all theoretical traditions or conceptual categories aligned to conservative or progressive thinking can be used to make sense of realities in India.

The left's failure to theorize caste in India has, of course, constituted one of its greatest failures, given evidence of the continuing role of caste in political mobilizations as a marker of identity, as a significant element in contemporary forms of Hindu nationalism, the prevalence of caste movements for equality, and other reinventions of caste within twenty-first century India. Drèze and Sen's book, *An Uncertain Glory*, raises questions related to the contradictions in contemporary India, a country in which the disparities between wealth and poverty are increasing, not decreasing.[35] Their critique of economic growth in India has been countered by a variety of other economists for whom per capita income is more important than people's basic access to essential resources. Yet, the contradictions that Drèze and Sen have highlighted cannot be wished away or ignored just because they do not fit well into the current mythos of growth and progress. They highlight the persistence and deepening of caste, class, and gender stratifications, where:

> caste stratification often reinforced class inequality, giving it a resilience that is harder to conquer ... . Gender inequality too is exceptionally high in India. It is the mutual reinforcement of severe inequalities of different kinds that creates an extremely oppressive social system, where those at the bottom of these multiple layers of disadvantage live in conditions of extreme disempowerment.[36]

David Harvey's book, *The New Imperialism*, was written at the cusp of the second war in Iraq and is an account of Pax Americana.[37] It is an account of the USA's global oil strategy, of the manner in which it has used its hegemonic power to expand a new imperialism via both the market and, when that has not been possible, through the means of war. In this light, this chapter provides an introduction to how accumulation

---

[35] Drèze, J. and A. Sen. 2013. *An Uncertain Glory: India and Its Contradictions*. Princeton: Princeton University Press.

[36] Drèze and Sen. *An Uncertain Glory*, p. 213.

[37] Harvey, D. 2003. *The New Imperialism*. Oxford and New York: Oxford University Press.

occurs through extra-economic means, in which the State plays a primary role in creating the conditions for accumulation. Harvey engages with Rosa Luxembourg's view of capital accumulation's 'dual character', that is, accumulation as a purely economic process and accumulation as a result of relations between capitalism and non-capitalist modes of production.[38] He also builds on Marx's views on 'primitive accumulation'—'the commodification and privatization of land and the forceful expulsion of peasant populations; the conversion of various forms of property rights (common, collective, state, etc.) into exclusive private property; the suppression of rights to the commons'—to describe the presence and persistence of these earlier forms of accumulation in the contemporary era.[39] He refers to the over-accumulation of capital and its investment in state-sponsored devalued assets, such as land, that, in turn, are bought back by private companies for much less than the original market price. However, this process of accumulation is not just about the possession of land as an asset but also entails the destruction of ways of life and its cultural and social sources. What makes ABD deeply destructive is that, apart from often paltry compensation packages and access to degraded land, it does not offer means for the cultural and social regeneration of affected communities. It is also a process that is 'both contingent and haphazard',[40] meaning that a number of factors outside of any normative framework are germane to ABD.

'The original sin of simple robbery', a phrase used by Hannah Arendt to describe the basis for original capitalist accumulation, has, as Harvey has observed, become robbery on a large scale as capitalism has expanded its footprint and commodified hitherto uncommodified sectors.[41] In this respect, it is clear that David Harvey and Karl Polanyi share a critical stance in regards to utopia of the self-regulating market economy. Both highlight and question the relationship of the state with the market and the 'double movement'—globalisation countered by anti-globalisation movements, and state-based welfare that provides a corrective to a pure market economy. There are, however, key differences, including the fact that Harvey explicitly deals with the consequences of the commodification

---

[38] Harvey. *The New Imperialism*, p. 137.
[39] Harvey. *The New Imperialism*, p. 145.
[40] Harvey. *The New Imperialism*, p. 149.
[41] Harvey. *The New Imperialism*, p. 144.

of land and labour, while Polanyi, who believes in the 'embeddedness' of the economy in the social, argues that such assets should not be commodified precisely because they are linked to life and livelihood and to values that are anti-commodity.[42] Polanyi famously describes land, labour, and money as fictitious commodities, the meanings of which have been socially constructed and manufactured and are crucially based on a denial of nature that we have inherited and that ought to be considered the patrimony of humankind. Labour's resistance of commodification is natural precisely because, as Gareth Dale, in an article on Polanyi, describes 'labour is inseparable from the human beings of which society consist[s], and land is their natural habitat, their insertion as fictitious commodities into the market mechanism' resulted in the subjugation of the social to the laws of the market.[43]

Polanyi's account of the complex consequences of the enclosures in the making of the English working classes, including the enactment of the Poor Law that led to dependency on the state and to furthering disengagements from the social, can be applied to an understanding of the role of the contemporary Indian state as Patron. The Indian government as the patron state has forced entire communities to a life of dislocation, of subsistence dependency. Arguably, this is a by-product of its commitment to a brave, new India built on countrywide connectivity and smart cities built on a vision of transnational urbanism. Polanyi engages with the nature and consequences of 'improvement' and growth and the role played by an earlier generation of technologies in transforming the means and relations of production. 'The Industrial revolution was merely the beginnings of a revolution as extreme and radical as ever inflamed the minds of sectarians, but the new creed was utterly materialistic and believed that all human problems could be resolved given an unlimited amount of material commodities.'[44] Substitute 'information revolution' in place of 'Industrial revolution' and what one sees is the very same belief in technologies resolving deep contradictions in society. Importantly, the

---

[42] Polanyi. *The Great Transformation: The Political and Economic Origins of Our Time*, p. 47.

[43] Dale, G. 2019. 'Social Democracy, Embeddedness and Decommodification: On the Conceptual Innovations and Intellectual affiliations of Karl Polanyi', *New Political Economy* 15(3): 381.

[44] Polanyi. 1944, p. 40.

question of the role played by the State in India in advancing, shaping, or curbing the framework for an embedded market economy, along with the role played by civil society and locality in embedding the market in traditions of sociality (the double movement), is the key to understanding the making of this new India.

The Indian State and Accumulation by Dispossession

What we are now seeing in India is the State support for multiple infrastructure projects designed to support the interests of the IT sector (and vice versa). This is one aspect of a larger attempt to ensure that the writ of the market is both extensive and expansive across multiple sectors. Munster and Strumpell highlight areas where a flexible, digital, transnational informational capitalism has begun to impact: 'Mobility, security, ethnicity, heritage, water, carbon-storing capacities of forests and "fair" trade relations have become incorporated into market relations.'[45] These are the new arenas where the State's ideational and material interests coincide. Whether we are seeing the rise of a hegemonic State in India that is wholeheartedly supportive of the hegemonic interests of big business remains debateable. The various counter-movements are heterogenous and, as is described in the chapters in this book, are both organized and unorganized. The attempt to restrict Free Basics was organized and involved movements committed to 'net neutrality' and the democratization of access for all Indian citizens, while resistance to the dominant digital in the informal economy involves an appropriation of the digital copy, as is the case with 'seed' and other commodities that have been digitally mastered. Nonetheless, as things stand, the State in India is increasingly difficult to distinguish from the market given the many mutual correspondences and the crucial role played by the State in creating the preconditions for a market economy. Sarbeswar Sahoo has argued that the 'Great Transformation' towards a new liberal market economy in India has been accompanied by declining investments in social welfare, resulting in an increase in poverty among sections of society such as the *Adivasi*s

[45] Munster, D. and C. Strumpell. 2014. 'The Anthropology of Neo-liberal India: An Introduction', *Contributions to Indian Sociology* 48(1): 1–16.

(tribals); growth in unemployment; disinvestments in public health; in unevenness and inequality; and counter-movements, such as that of tribals, against projects based on extraction, like mining (Vedanta) in Odisha and Maoist movements in east and central India.[46] This book tries to capture this 'double movement' and the role of a variety of actors—the State, civil society, and locality—in trying to curb ABD.

There are clear correspondences between India's external strategy and its internal policies. In the context of multilateral and bilateral trade within a neo-liberal economic framework, there is a strong commitment to the liberalization of each productive sector. It is this progressive liberalization that has inevitably led to major confrontations between the State and its people. The Indian state's active role in acting as a handmaiden to big businesses, exemplified in its commitments to infrastructure support, tax breaks, and its role as broker in the acquisition of land, has not been popular with large sections of the population, especially those who are dependent on agricultural production as their means of livelihood. Land in India is a premium asset and its commodification through recourse to legal mechanisms, such as 'eminent domain', remains a critical issue in contemporary India. Large tracts of both common and private land have been possessed by the State for infrastructure projects, SEZs, and IT parks and corridors. Michael Levien, in an article on ABD and SEZs in India, refers to the Indian state as a 'land broker state' that has approved 581 SEZs since 2005.[47]

Levien cites the example of Mahindra World City, a joint venture between Mahindra Lifespace Developers, a subsidiary of the Indian multinational company Mahindra and Mahindra, and the public-sector based Rajasthan Industrial Development and Investment Corporation. This 3,000-acre SEZ is ex-farmland and the Mahindras have been able to make major profits by acquiring low-priced 'dispossessed land' that, in turn, has been sold for 'market' prices to commercial and residential

[46] Sahoo, S. 2017. 'Market Liberalism, Marginalised Citizens and Countermovements in India', *Asian Studies Review* 41(1): 1–19.

[47] Levien, M. 2011. 'Special Economic Zones and Accumulation by Dispossession in India', *Journal of Agrarian Change* 11(4): 461, 454. See also Jenkins, R. 2011. 'The Politics of India's Special Economic Zones', in S. Ruparelia, S. Reddy, J. Harriss, and S. Corbridge (eds), *Understanding India's New Political Economy: A Great Transformation?* London and New York: Routledge, pp. 49–65.

buyers.[48] Such examples are all too common throughout India. The issues related to real estate and the IT sector were also graphically illustrated by the rise and fall of an iconic software company, Satyam Computers, in late 2008. Its owners, the Raju brothers, were deeply involved in real estate and infrastructure projects. The tens of thousands of acres owned by the Rajus via their real estate business, Maytas Properties was the focus for investigation by the Government of India—a direct consequence of the revenue scam that affected Satyam Computers in early 2009. As Venkateshwarlu has observed:

> It remains a mystery how Raju managed to acquire 6,800 acres of land, with preliminary reports suggesting benami (assumed names) purchases or the floating of a number of companies to circumvent laws ... revenue officials are looking at reports that he had bought up large chunks of lands originally assigned by the government to the Scheduled Castes and Scheduled Tribes, some in connivance with a local Congress leader in Ranga Reddy district neighbouring Hyderabad.[49]

Such acquisitions of land by the State for IT companies is commonplace in India.

David Harvey characterizes ABD as the dominant form of accumulation under neo-liberal capitalism, a dispossession that is based on privatization, financialization, the globalization of multilateral trade regimes, the conversion of the world's commons into private property with the help of the State, and the extraction of rents through intellectual property (IP). The State, in its new avatar, uses coercive means and extra-economic power to transfer land and other resources from the agricultural, peasant populations to the MNCs, which would otherwise have not been able to access such resources.[50] This process is backed up by finance: 'The umbilical cord that ties together accumulation by

---

[48] Levien. 'Special Economic Zones and Accumulation by Dispossession in India', p. 459.

[49] Venkateshwarlu, K. 2009. 'Maytas Twins', *Frontline* 26(3), 31 January–13 February, available at http://www.frontline.in/static/html/fl2603/stories/20090213260301000.htm; accessed on 11 May 2018.

[50] See Levien, M. 2012. 'The Land Question: Special Economic Zones and the Political Economy of Dispossession in India', *The Journal of Peasant Studies* 39(3–4): 933–69.

dispossession and expanded reproduction is that given by financial capital and the institutions of credit, backed, as ever, by state powers [*sic*]:[51] To Harvey, ABD is financed by over-accumulated capital in the global economy, although others, such as Parthasarathy, have argued that the source of this capital is also local in origin and reflects liquidity among the richer peasant castes: 'The rise of Hyderabad city in south India as one of India's key information technology and software centres needs to be explained as much by its links to global capital, as by the flow of provincial capital from coastal Andhra, especially through the landowning dominant Kamma caste rich peasants.'[52] These caste/class groups act as rentiers who purchase such lands, often with the help of the State. They alter these lands into SEZs that are, in turn, sold to the MNCs.

The extraordinary scale of such projects is best illustrated by Amaravati, the new capital city project in the southern Indian state of AP. Amaravati, modelled on Singapore, is being built on 40,000 acres of both common and prime agricultural land that has been taken possession of by the State. These lands were originally farmed by both rich absentee landlords and poor tenant farmers. Compensation for the loss of land has, however, been inflected by class/caste compulsions. While absentee landlords have been allocated real estate in the new city, the poor tenant farmers and landless labourers have been offered a paltry monthly sum of Rs 2,500. The plan is to create a gleaming smart city that is exclusive and in which there is little or no space for those who will provide services to this city. As an avowed technocrat, this is the second occasion when the chief minister has directed the development of a global enclave, the first being the Hitech City in Hyderabad, for which land was acquired through application of the Urban Land Ceiling Act and the Land Acquisition Act (thereby dispossessing small farmers), and was then given away for below market rates to international software giants, such as Microsoft, and national firms, such as Wipro, Tata Consultancy Services (TCS), Satyam, and other software companies.[53] The Marxian media critic Padmaja Shaw, in a personal email, stated that:

[51] Harvey, *The New Imperialism*, p. 152.

[52] Parthasarathy, D. 2015. 'The Poverty of (Marxist) Theory: Peasant Classes, Provincial Capital, and the Critique of Globalisation in India', *Journal of Social History* 48(4): 824.

[53] Pakalapti, U.V. 2010. 'Hi-tech Hyderabad and the Urban Poor: Reformed Out of the System', in S. Banerjee-Guha (ed.), *Accumulation by Dispossession: Transformative Cities in the New Global Order*. New Delhi: SAGE, pp. 125–50.

[It] is a massive accumulation by dispossession case where the state is acting as a predatory real estate agent. They have been burning crops, slapping cases on farmers who refuse to give up their land. The ministers in Chief Minister Chandrababu Naidu's cabinet are widely regarded as a mafia. This whole thing can turn out to be his political nemesis, even as it destroys one of the most prosperous and settled farming community that also gives a lot of livelihood for the poor around the area.[54]

Observations on the IT Compact: India and the USA

The Indian state's negotiations within the geopolitics of the information economy have varied over the seven decades since Independence. While the immediate post-Independence period was characterized by dependency on the USA and the UK for computing and telecommunication technologies, the 1970s were watershed years where the Indian state made a determined bid to strengthen its own information capacities. One of the most striking events in this era was IBM's withdrawal from India in 1977 after it refused to comply with the demand to dilute its ownership to include Indian nationals and provide advanced computing technologies.[55] IBM's exit led to the strengthening of the remit of Electronics Corporation of India (ECIL) and Indian Telephone Industries, as well as investments in State capacity in IT and telecommunications. Despite occasionally frosty relationships with the USA, there were also other moments characterized by close collaboration, such as in the area of satellites and the Satellite Instructional Television Experiment (SITE). While space research began in India in 1962, perhaps the most significant state-based attempt to expand communication was the SITE that was conducted between 1975 and 1976. This attempt at satellite-based distance education was among the most ambitious of its kind and involved 2,330 villages in 6 states. While the satellite used was on loan from the US-based National Aeronautics and Space Association (NASA), the entire complex of

---

55 See Grieco, J.M. 1984. *Between Dependency and Autonomy: India's Experience with the International Computer Industry*. Berkeley, Los Angeles, and London: University of California Press; and Subramanian, C.R. 1992. *India and the Computer: A Study of Planned Development*. New Delhi: Oxford University Press.

ground-level infrastructure was created and maintained through Indian effort and the software produced locally.

The era of liberalization since 1991 has seen India developing a much closer relationship with the USA than in any other period, with crucial relationship developing around information technologies in particular. Nonetheless, Bajpai suggests that even in the context of this new relationship with the USA, 'India has not altogether lost its fear of US imperialism.'[56] For its part, it has been argued that the USA's cooperation with India in the knowledge sector is critically and primarily about access to its talent pool and only secondarily about technology transfers.[57] The issue of advanced technology transfers between the USA and India has always been fraught. In 1965, for example, the Indian government had approached the USA for Scout rocket technology but was denied the same as the USA felt that it could be used to develop India's ballistic missile capabilities.[58] Similarly, in the 1980s, India's request for Cray supercomputing technology was initially refused as it was felt that such 'dual-use' technologies could fall into the hands of the USSR; in the latter-half of the Cold War era, the USSR became a more dependable, regular partner for India in the supply of such technologies. Nonetheless, as Raju has observed, it was an MoU between India and the USA, signed in 1984, that led to sales of 'Cray XMP-14 supercomputer for India's Meteorological Department and the advanced 'Silicon-on-Saffire" microprocessor chip for India's INSAT-2 satellite.'[59]

While India is firmly embedded within the global digital economy today, it is still dependent on the US market for its software exports (between 60 and 80 per cent). India's largest IT services exporter, TCS, had revenues of Rs 37,325 crore (US$5.6 billion) and a net profit of Rs 8,716 crore (US$1.3 billion) in 2010–11. Only 10 per cent of its revenues came from the domestic market and the rest was earned through

[56] Bajpai, K. 2015. 'Five Approaches to the Study of Indian Foreign Policy', in D.M. Malone, C. Raja Mohan, and S. Raghavan (eds), *The Oxford Handbook of Indian Foreign Policy*. Oxford: Oxford University Press, p. 25.

[57] See Sikri, R. 2009. *Challenge and Strategy: Rethinking India's Foreign Policy*. New Delhi: SAGE.

[58] Reddy, S.V. 2011. 'India and Outer Space', in David Scott (ed.), *Handbook of India's International Relations*. London and New York: Routledge, pp. 311–20.

[59] Raju, G.C.T. 1990. 'U.S. Transfers of "Dual-Use" Technologies to India', *Asian Survey* 30(9): 840.

its exports of software and consultancy services mainly to the USA and Europe.[60] Further, as India's policies related to the Internet-based 'new' media economy are not clearly defined, this has led to controversial and as yet unresolved issues related to IP; taxation of cross-border software licensing transactions; lack of distinct e-commerce legislations; and governance of the Internet.[61] While the strategic importance of India to the USA has been fortified in the context of China's regional ambitions, India continues to remain the dependent partner in this relationship.[62] However, in spite of this dependency, local compulsions can and do make a difference in its geopolitical strategies. While the incumbent government is pro-business and pro-trade, its Hindu nationalist agenda includes support for the cause of Greater India along with digital independence and information sovereignty. This rather paradoxical stance has in part been fuelled by the present BJP government's strongest partner, the RSS, a Hindu right-wing movement, whose members include the present prime minister and most of the cabinet.

The Indian pharmaceutical industry's role in manufacturing generics and government support for this sector and compulsory licensing has been a consistent issue for the US firms over the years. Seemantani Sharma has observed:

> the lobbyists have played the platitudinous trumpet of an unpredictable compulsory license regime, the non-conformity of the patentability criteria under Section 3(d) of the Indian Patent Act, 1970 (the Patents Act) to TRIPS—the trade-related intellectual property rights regime of the World Trade Organisation—the absence of a trade secret and regulatory test data protection regime, backlog of patent and trademark applications and a dilatory patent enforcement process.[63]

---

[60] BW Online Bureau. 2014. 'A Quiet Revolution', *BusinessWorld*, 8 November, available at: http://www.businessworld.in/article/A-Quiet-Revolution/08-11-2014-63046/; accessed on 12 March 2019.

[61] Goradia, S. and C.P. Tello. 2014. 'Conflicts and Issues under the US–Indian Tax Treaty', *International Journal of Taxation* 10: 27–31.

[62] Wagner, C. 2010. 'India's Gradual Rise', *Politics* 30(S1): 63–70.

[63] Sharma, S. 2017. 'Despite Modi's New IPR policy, US Continues to Cry Foul over India's Laws', *The Wire*, 31 March, available at https://thewire.in/120122/national-ipr-policy-cautionary-pessimism-continues-washington-d-c/; accessed on 17 May 2018.

India's contraventions of both copyright and big pharma patents have been especially contentious issues. The Section 301, 2014, executive summary highlights continuing violations related to copyright, inclusive of online and video piracy, although its key concern relates to patents and regulatory data protection. The pressure to place India in this category came from a trade lobby, the Alliance for Fair Trade with India (AFTI), co-chaired by National Association for Manufacturers (NAM) and the membership of the U.S. Chamber of Commerce's Global Intellectual Property Center, including CropLife America, Pharmaceutical Research and Manufacturers of America, Telecommunications Industry Association, National Foreign Trade Council, the Motion Picture Association of America, and the Biotechnology Innovation Organization.[64] While the AFTI has complained about India's violations of copyright, data protection, and trade secrets, its key point of contention has been with the Indian government's commitment to the compulsory licensing of drugs and the generic manufacture of drugs that are deemed to be in the national interest.

India's stance at WIPO does suggest that it is difficult to assess the country's commitment to IP in black-and-white terms. I have described the State in India as an ambivalent entity caught as it is between the imperatives of neo-liberalism and its commitments to the welfare of its publics, particularly the millions of people who rely on the State for their very survival.[65] The first draft of a National IP Rights Policy was released in December 2014, but whether all this will lead to the 'strengthening' of the IP writ of global and local companies in India is debateable. The decision taken by Monsanto in 2016 to withdraw its new GM cotton seed from India in protest against the government pressurizing it to share its proprietary technology with local seed companies does suggest that the GoI continues to be involved in playing an elaborate IP game that is based on both openness and closure. The SJM has strongly objected to the US pressure to dilute India's patent laws in favour of global pharma companies. Bagchi, writing in the *Hindu*, has reported:

---

[64] USTR. 2014. 'Business Coalition Seeks PFC Designation for India in Special 301 Report', 14 February, *Inside US Trade* 32(7): 1–63, available at https://ustr.gov/sites/default/files/USTR%202014%20Special%20301%20Report%20to%20Congress%20FINAL.pdf.

[65] Thomas. 'The Ambivalent State and the Media in India'.

The SJM has ... objected to the policy amendments in 'compulsory licensing' especially while granting permission to NATCO for manufacturing cancer drugs which [were] being sold by BAYER of Germany 'at exorbitant prices' in the country ... red-flagged 'illegitimate demand of data exclusivity on pharmaceuticals' whereby the Drug Regulatory Authority of India will be prohibited to disclose trial results to the Indian generic companies. SJM has ... flagged a whole lot of other areas where US is trying to mount 'pressure'. ... According to the RSS outfit, the government should demand 'protection' for products like Darjeeling tea, Basmati rice, textile goods and several other agricultural products which have its origin in India.[66]

While the need for stronger IP protection for US cultural industries in India continues to be an aspect of the geopolitics of information, for the most part, there have been attempts to forge closer ties between these two countries, not only in the matter of trade but also in terms of security, the fight against global terrorism, and in the celebration of commonly shared values related to democracy and the rule of law. The displacement of Pakistan as an ally of the USA has resulted in a closer relationship with India: a role that has become more pronounced in the context of China's economic and territorial ambitions in South Asia. Although India continues to import arms from Russia, the USA is now a key military ally. India has become the biggest arms importer in the world and the biggest buyer of arms from the USA (worth $1.9 billion in 2013),[67] and is involved in joint military exercises and the forging of military policy in the Indian Ocean region. Significant shared issues are the growing influence of China, its territorial ambitions, and ensuring maritime security and 'commons' in the Indian Ocean region. Thus, the Indo-Pacific region has become the setting for a geopolitical alignment between the USA and India, in which India has been singled out as the 'net security provider' in the region. While the USA would like India to become its official ally, the Indian government has consistently opted for a relationship based

---

[66] Bagchi, S. 2015. 'US "Threats" to Patent Act irks RSS Outfit', *The Hindu*, 23 January, available at http://www.thehindu.com/news/national/us-threats-to-patent-act-irks-rss-outfit/article6815593.ece; accessed on 22 May 2018.

[67] Plimmer, G. and V. Mallet. 2014. 'India Becomes Biggest Foreign Buyer of US Weapons', *Financial Times*, 24 February, available at http://www.ft.com/intl/cms/s/0/ded3be9a-9c81-11e3-b535-00144feab7de.html#axzz3e7k2WjIq; accessed on 11 May 2018.

on preserving its 'strategic autonomy'—thereby ensuring its freedom to exercise maximum options in the context of its dealings and actions with the outside world.[68] It remains unclear, however, to what extent the Indian government will be able to preserve this autonomy and forge an independent position on its national and international economic priorities.

During the same period, India's opening up to the global market, trade, and capital has been the defining characteristic of its contemporary political economy, and its emergence as a global hub for software and allied services has undoubtedly defined its strategic relations with the USA and other countries. The vision of India as an IT superpower has been embraced by its technocrati, although the State has found it a lot more difficult to communicate this vision to the masses who have other, more basic, priorities and who continue to rely on the State for basic services including food, employment, and shelter. The economic vision of a digital superpower has become key to India's national identity: a vision that has been highlighted in the incumbent government's framework for Digital India. Wyatt has pointed out that this imagined national economy is the basis for a nationalist ideology: 'a basis for citizenship, mediating economic "truth" and engineering consent for economic practice.'[69] While the digital economy is the basis for the most extensive capital accumulations ever seen in the history of capitalism on the one hand, it is also proving to be difficult to manage and control on the other, in spite of millions of dollars of investments in encryption, enforcement, and enclosure. This is not only a story of contestations between IP industries and an assorted menagerie of cultural pirates but is also about ongoing negotiations between dominant IP countries, especially the USA, and 'recalcitrant' countries, such as India. Some of India's actions, such as changes to its patent laws, have led to a tightening of the proprietorial digital economy, while others, such as the habitual release of digital 'pirates' by the lower courts in India and support for free and open source software (FOSS)-based public sector software in some states in India, suggest that the Indian government's stance on the

---

[68] Upadhyay, S. n.d. 'The Indo-Pacific and the Indo-US Relations: Geopolitics of Cooperation', Institute of Peace and Conflict Studies, available at http://www.ipcs.org/issue-brief/china/the-indo-pacificnbsp-amp-the-indo-us-relations-geopolitics-of-256.html; accessed on 25 May 2018.

[69] Wyatt, A. 2005. '(Re)imagining the Indian (Inter)national Economy', *New Political Economy* 10(2): 163–79.

digital is also motivated by its desire to hold on to some 'sovereignty' in its engagements with the knowledge economy.

The State and the Digital

I have begun by canvassing India's place within the geopolitics of information and by questioning the relationship between the State and the digital. One key dimension to this relationship is the State as a pan-Indian entity and its multiple engagements with the digital, in which it variously acts as an initiator of countrywide projects such as Aadhaar, is involved in the production of both hardware and software, is the regulator, and is also involved in its own right in shaping the digital market. Increasingly, these entanglements need to be seen against the State's role as an important player in and arbiter of the emerging transactional economy in India.[70] The chapters that follow explore the politics of information, in particular the ambiguities and contestations of the global–local in the evolving knowledge economies in India, both formal and informal. In doing so, we are able to explore how global rules and projects are filtered through the prism of the Indian government and how these are shaped, in turn, by multiple ambivalences. The algorithmic moment may have arrived in India, but how algorithms are used by the State for 'biopolitical' (surveillance, monitoring, disciplining, and control) purposes and their uses by the market for precise consumer marketing are two sides of a multifaceted story. This story is also about the informalizing of the information economy and civil society initiatives directed towards claiming access for all. While it may be difficult to predict how the knowledge economy will evolve in India, there is no doubting the fact that it will be shaped by multiple actors. The case studies presented in this book constitute a necessary exploration of this terrain, by providing some entry points that help us to understand the political dimensions of the digital in India.

The structure of this book is as follows. Following this 'Introduction', Section I includes three chapters that highlight various aspects of the 'Control State'. It includes Chapter 1, 'The Expansion of Politics as Control: Surveillance in India'; Chapter 2, 'Leisure, Surveillance, and the

---

[70] Athique, A. and E. Baulch (eds). 2019. *Digital Transactions in Asia: Economic, Informational and Social Exchanges.* New York: Routledge.

Private Sector in India'; and Chapter 3, 'Software Patent Manoeuvres' dealing with the State's attempts to expand its digital footprint through its many policy and programmatic commitments to surveillance, PPPs in expanding State control, and its support for software patents. This is followed by Section II that is on the contested nature of India as a sovereign/ambivalent State, caught as it is between local compulsions and transnational pressure. Chapter 4 is on 'Digital (Transgenic) Seed and Its Copy', while Chapter 5 explores 'The Politics and Geopolitics of Internet Governance'. Chapter 6, 'The WIPO Treaty for the Visually Impaired as a Double Movement', explores a treaty that was preceded by the Indian government's amendments to the Copyright Act supportive of the interests of India's many million visually impaired citizens. The 'Conclusion' on India's 'strategic autonomy' and the politics and geopolitics of digital India attempts to explore the Indian state's 'strategic autonomy' on matters related to the digital and the complex relationship that it has with a variety of global and local issues related to the geopolitics of information. It argues that the shaping of India's digital futures will be multifaceted and involve both external pressures and internal recalibrations.

Section I

The Control State

1

The Expansion of Politics as Control

Surveillance in India

Arguably, the processes associated with ABD include the many ways in which people's sovereignty and freedoms are being curtailed through State and private interferences in their lives. In the absence of a secure privacy policy in India, there is ample scope for the State to intimately control people's lives either through projects such as the UID initiative or through the legitimization of forms of sousveillance. Given that a major focus of surveillance is precisely those communities who are part of the precariat and/or who have been dispossessed of their lands and livelihoods, or who are already living in poverty, such as indigenous communities in Maoist-held regions in east and central India, surveillance is a form of biopolitics and an exercise of power directed towards those communities that are resisting the privatization of their lands and livelihoods. There is also a sense in which surveillance is a corollary of the Indian State's embrace of accumulation in the context of neo-liberal growth, best exemplified by the fact that the boom in the surveillance industry has been an outcome of the growth of the property market, IT corridors, gated communities, and smart cities. It is indeed ironic that though many of these have been built on land forcibly taken from the poor, these very people are now unwelcome in these new citadels of power. In other words, 'financialization', which arguably has contributed to neo-liberal growth in India, has played its part in the accentuation of surveillance. Harvey has argued in the case of the USA that the cost of its imperialism 'has to be paid at home in terms of civil liberties, rights and general freedoms'.[1] In India, it would seem that the

[1] Harvey, D. 2004. 'The "New" Imperialism: Accumulation by Dispossession', *Socialist Register* 40: 82.

State's attempts to control its minorities and other recalcitrant groups are tied to the expansion of surveillance and curtailment of personal freedoms.

The surveillance of populations in India became a priority public policy after threats to its national security, including events such as the 2001 attack on the Indian Parliament and the Mumbai terrorist attacks in 2008. The 2001 attack led to the Prevention of Terrorism Act (POTA), which that has been likened to the Patriot Act in the USA.[2] While the desire to track the movements of terrorists is a ubiquitous aspect of security regimes throughout the world, the Indian government has also invested in schemes such as the UID project that is, at least in official parlance, a digitized identity (ID) card that would allow for greater transparency in the provision of welfare entitlements, though it can also be used to track and monitor the whereabouts and behaviours of individuals. This ability to monitor suspect populations is especially problematic in the context of the present government's Hindu nationalist credentials and public pronouncements by both sitting politicians and ideologues linked to the family of right-wing Hindu organizations that espouse Hindutva as a national ideology and that have taken issue with minorities, inclusive of Muslims, Christians, and Sikhs. They have perfected and mainstreamed a counter-discourse against secularists, rationalists, and anyone else who is critical of the attempts to rewrite the history of India in line with Hindu India, and have used online platforms to attack and vilify those who are critical of the project of Hindu India. That the government in power is sympathetic to Hindu right-wing groups is a cause for worry, as some like the RSS, the Bajrang Dal, and the Vishva Hindu Parishad (VHP) have fascist leanings. Moreover, successive governments over the last decade have invested in a range of databases that capture specific facets of data on India's population specifically for surveillance purposes, with the Central Monitoring System (CMS) that functions without any judicial oversight being a key node in the government's surveillance apparatus.

Do such investments in surveillance make India any different from other countries such as the USA? Perhaps not, although I think the key difference is the underlying majoritarian politics and the public threats to surveil minority groups. This chapter begins with a brief introduction to

---

[2] Krishnan, J.K. 2004. 'India's "Patriot Act": POTA and the Impact on Civil Liberties in the World's Largest Democracy', *Law and Inequality* 22(2): 265–300.

the theorization of surveillance, followed by sections on the participatory nature of surveillance, surveillance as practice, history of surveillance in India, surveillance partners, surveillance and privacy, and the UID initiative, before concluding with a section on surveillance in the context of e-governance.

Theorizing Surveillance

The political economist Armand Mattelart has written an intriguing account of the history of surveillance in Western societies: from Jeremy Bentham's (1748–1832) Panopticon to Franz Josef Gall's (1758–1828) cranioscopy that assessed craniums and helped predict 27 faculties including the 'propensity to murder', to Gustave Le Bon's (1841–1931) study of skulls that reinforced his antipathy to racial intermixing, to surveillance during the Cold War, to a contemporary society that is based on continuous surveillance.[3] As he observes: 'In this generalized control society, governed by the managerial model, the ability to anticipate individual behavior, identify the probability of a specific behavior and construct categories based on statistical frequency is the common thread among the "styles" of marketing specialists, the "scores" of financers and the "profiles" of the police.'[4] What makes surveillance an especially interesting area from a political economy perspective is the universalizing of PPPs in the making of the technologies of surveillance, public-sector subsidization of such initiatives, and, in the absence of clear jurisprudence supportive of privacy, public–private sharing of the fruits of dataveillance. An example of a public–private surveillance project is the US-based Cell All System,[5] in which the project partners are NASA and two private sector firms, namely, Qualcomm and Synkera Technologies. While NASA's expertise is in the area of developing chemical sensing in low-powered platforms, Qualcomm is involved in developing smartphone apps and network software, while Synkera Technologies will develop a technological nose

[3] Mattelart, A. 2010. *The Globalization of Surveillance: The Origin of the Securitarian Order*. Cambridge and Malden: Polity Press.

[4] Mattelart. *The Globalization of Surveillance*, p. 184.

[5] US Department of Homeland Security, 'Privacy Impact Assessment for the Cell All Demonstrations', 20 March 2011, available at https://www.dhs.gov/xlibrary/assets/privacy/privacy_pia_s&t_cell_all.pdf; accessed on 22 May 2018.

that will sniff out explosives, toxic industrial chemicals, chemical warfare agents, and nerve agents.

In the context of India, where there have been major public–private investments in a variety of national and state-led surveillance projects, private sector firms including Alcatel–Lucent India and Siemens Information Systems along with government-owned firms such as Telecommunications Consultants India are among more than 70 companies involved in supplying monitoring equipment.[6] One needs to acknowledge that India does not have any privacy legislations to speak of. The Associated Chambers of Commerce and Industry of India (ASSOCHAM) along with PricewaterhouseCoopers (PwC) released a report in 2013 that revealed systematic investments in phased city surveillance in India and the availability of funds for a variety of projects related to surveillance systems and equipment, video analytics, network connectivity, and cyber patrols, among other initiatives.[7] Monahan and Mokos, commenting on the privatizing of government functions in the context of public–private neo-liberal partnerships related to the Cell All System, highlight the privatization of risk, and therefore surveillance, that is a consequence of risk propaganda: 'a certain type of participatory surveillance must be cultivated through discursive appeals to individuals—whether patriotic duty to avert mass casualty disasters, personal interest to save oneself or one's loved ones from carbon monoxide poisoning, or individual desire to be part of an innovative technological project.'[8] In other words, both sousveillance and surveillance are aspects of the State's approach to the monitoring of its civilian populations.

[6] See Xynou, M. 'The Surveillance Industry in India: At Least 76 Companies Aiding Our Watchers!', The Centre for Internet & Society (CIS), 2 May 2013, available at http://cis-india.org/internet-governance/blog/the-surveillance-industry-in-india-at-least-76-companies-aiding-our-watchers; accessed on 11 May 2018.

[7] ASSOCHAM and PwC, *Safe Cities: The Indian Story*, 2013, available at http://www.pwc.in/en_IN/in/assets/pdfs/industries/government/safe-cities-the-india-story.pdf; accessed on 21 May 2018.

[8] Monahan, T. and J.T. Mokos. 2013. 'Crowdsourcing Urban Surveillance: The Development of Homeland Security Markets for Environmental Sensor Networks', *Geoforum* 49: 280.

The development of surveillance studies in the wake of key writings by David Lyon,[9] Mark Andrejevic,[10] and others, along with specialist journals such as *Surveillance & Society*, is indicative of growing interest in academic explorations of surveillance in the context of the twenty-first century. The trajectory of surveillance studies has changed from the abiding interest in the disciplining of individuals through apparatuses such as the Panopticon to surveillance in control societies where it has become a routine aspect of everyday life. Mattelart makes a distinction between the disciplinary society based on the exercise of centripetal power and the security society where power is centrifugal, de-territorialized, constant, and intimate, and in which whole populations are subject to constant surveillance.[11] As Wood describes it, surveillance is found wherever there is 'purposeful, routine, systematic and focused attention paid to personal details for the sake of control, entitlement management, influence or protection'.[12]

The mining of personal details and the focus on the body as the site for information gathering characterizes contemporary forms of biopolitics directed towards a sifting and sorting out of individuals by the State and private corporations, both separately and at times together. What distinguishes the present surveillance regime is the routinization of surveillance in everyday life and its use as a form of population management and consumer control. While anti-terrorism-related surveillance stands at one end of the typology, at the other end is the altogether routine type of data gathering that is exercised in the context of a great variety of mundane activities that individuals are involved in, in the context of informational capitalism. These include: using our credit cards to purchase a variety of

---

[9] Lyon, D. 1994. *The Electronic Eye: The Rise of Surveillance Society*. Cambridge and Malden, MA: Polity Press and Blackwell; Lyon, D. 2001. *Surveillance Society: Monitoring Everyday Life*. Buckingham: Open University Press; and Lyon, D. 2004. 'Globalising Surveillance: Comparative and Sociological Perspectives', *International Sociology* 19(2): 135–49.

[10] Andrejevic, M. 2007. *iSpy: Surveillance and Power in the Interactive Era*. Lawrence, Kansas: University Press of Kansas.

[11] Mattelart. *The Globalization of Surveillance*, p. 9.

[12] Wood, D.M. (ed.). 2006. *A Report on the Surveillance Society*, for the Information Commissioner by the Surveillance Studies Network, p. 42, available at http://ico.org.uk/~/media/documents/library/Data_Protection/Practical_application/SURVEILLANCE_SOCIETY_FULL_REPORT_2006.PDF; accessed on 23 May 2018.

items; our everyday uses of social networking; and transactions that involve our use of the digital. In other words, it is when surveillance is seen in both its private and public manifestations—surveillance exercised at home and in the streets via closed-circuit television (CCTV) cameras, drones, and so on, when individuals are stationary or mobile, when they are at home or cross borders—that the extent of surveillance becomes real.

This constant expansion of surveillance has, of course, been best expressed by Foucault. In his classic book, *Discipline and Punish: The Birth of the Prison*, Foucault deals with the expansionary nature of surveillance and the fact that the exercise of power is constant—a reality that is also experienced in the context of contemporary control and security societies: 'The disciplinary power' is 'absolutely indiscreet, since it is everywhere and always alert, since by its very principle it leaves no zone of shade and constantly supervises the very individuals who are entrusted with the task of supervising; and absolutely "discreet", for it functions permanently and largely in silence.'[13] While Foucault's understandings of the relationship between biopolitics and modern forms of governmentality remain critical to our understanding of the exercise of surveillance, there have been arguments in favour of a post-Foucauldian understanding of surveillance in which the locus of attention is on the shift from biopolitics as a discourse to biopolitics that is now focused on the individual body as the source of information; in other words, biopolitics exercised through fingerprinting, iris scans, and the collection of deoxyribonucleic acid (DNA) as the means of sorting out populations. Furthermore, there is a lot more accent today on the labour that individuals expend on the generation of data that is of interest both to the State and to search engine giants, such as Google, that aggregate such data and make it available to advertising and other firms who are interested in an individual for a focused marketing of goods and services. Lemke suggests that 'we have to ask how strategies of power mobilize knowledge of life, and how processes of power generate and disseminate forms of knowledge. This perspective enables us to take into account structures of inequality, hierarchies of value and asymmetries that are (re)produced by biopolitical practices.'[14]

[13] Foucault, M. 1977. *Discipline and Punish: The Birth of the Prison*. New York: Pantheon Books, p. 177.

[14] Lemke, T. 2010. 'From State Biology to the Government of Life: Historical Dimensions and Contemporary Perspectives of "Biopolitics"', *Journal of Classical Sociology* 10(4): 433.

Interactivity, Participation, and Surveillance

Andrejevic has argued that key to the creation of digital enclosures today is the emphasis that has been given to the technologies of liberation, in particular, mobile phones and social networking sites. The emphasis is on the value of the freedoms that accompany the latest version of the iPhone, and specifically on agency that is a result of participation and interactivity. This domestication of the discourse of interactivity is an extraordinarily important means of expanding surveillance, given that there are greater correspondences between 'interactive' technologies, functions, and processes today which suggest that consumers have the sovereignty and independence to control the flows of information generated by themselves. Basing his work on reality television (TV) and programmes such as *American Idol*, Andrejevic has argued that the missing dimension in the celebrations of the 'active audience' is an understanding of the political economy of such offerings and our failure to consider the actual conditions under which this potential is being developed, namely, the dependence of new forms of interactivity on the economic priorities of large corporations dependent on the goodwill of government bureaucracies for access to lucrative markets'.[15] The freedom to participate and become interactive remains a key aspect of the regimes of truth that have incorporated this narrative into their myths of progress and growth. What we are also seeing in the context of this freeing of the individual from his/her shackles to the dominant state, is a narrative of the new state that is not as interested in the exercise of vertical power, but is convinced of the need to also become participatory, interactive, and involve citizens in decision making.

To be fair, there is a functional aspect to the making of Gov 2.0 given that informationalization has also contributed to investments in bureaucratic efficiency and to a rationalization of services. So, arguably, it is the global project of macro informationalization, along with the expansive project of consumer-oriented micro informationalization that provides both the rhetoric and discourse of participation and interactivity. However, while the concept of interactivity is often used in conjunction with new technologies, the domestication of participation has been an issue of concern among critical scholars involved in exploring communications for social change. Both concepts, nonetheless, place the individual at the forefront of social change. Over the last two decades, we

[15] Andrejevic. *iSpy*, p. 264.

have seen a massive increase in what one can call the 'industrialization of participation'—in terms of theories, processes, institutions, and a global validation that participation is the missing link in development at the level of intergovernmental agencies, government, the private sector, and civil society. However, this participation is increasingly the means by which consumers are becoming tied even closer to the logic of informational capitalism.

In another volume, Andrejevic has focused his attention on the consequences of information overload and the politics of managing large amounts of information, that is, big data.[16] He tries to make sense of the many new methods of symbolic and cultural mining that attempt to understand not only what we think and how we think, but also individual pattern-based predictions of how and what we will think of in the future—a boon for surveillance agencies, be it the State that is keen to predict how a terrorist will think so as to pre-empt him/her or the consumer and leisure industries that are keen to wed individual profiles to consumer ends. These include

> data mining and predictive analytics (which automate information processing and displace explanation with correlation); sentiment analysis (which purports to translate emotional response and individual opinions into machine readable data that can be mined); prediction markets (which replace credentialed expertise with aggregate demand and calls this wisdom); body language analysis (which privileges immediate bodily reactions over the vagaries of narrative content) and neuro marketing (a form of body language analysis that requires special equipment).[17]

In other words, the processes of biopolitics, the drilling and mining, have increasingly become related to understanding the nature of 'affect'. And this data, arguably, is directed towards creating possibilities for the manipulation of behaviours for the ends of both the State and the corporate sector.[18] Increasingly, the approach taken by security agencies is to collect whatever data is available through whichever means—the

---

[16] Andrejevic, M. 2013. *Infoglut: How Too Much Information is Changing the Way We Think and Know*. Hoboken: Taylor & Francis.

[17] Andrejevic. *Infoglut*, p. 4.

[18] See Esposti, J.D. 2014. 'When Big Data Meets Dataveillance: The Hidden Side of Analytics', *Surveillance & Society* 12(2): 209–25.

hope being that the collection of all data will eventually reveal patterns in individuals that require investigation.

Contemporary surveillance consists of population, security, and consumer-oriented watching that is more often than not carried out without the knowledge of citizens. In the context of what is a generalized war against terrorism, quite a few States have become national security states in which policies related to surveillance are hardly, if ever, debated, and in the absence of checks and balances, data collected by one agency are shared with another located in some other part of the world. Decisions on surveillance are taken by committees that operate beyond the oversight of Parliament, and the reason of security has become a dominant excuse for the extraordinary increases in surveillance budgets of governments throughout the world. What makes this issue complex is the fact that along with the common forms of surveillance, the profiling of players in financial markets and the financial health of nations are also objects of surveillance. As Wood has observed, surveillance is the basis for 'policy laundering' and 'both public agencies and private information companies collect, collate and perform algorithmic operations on data relating to nation-states to assess their credit worthiness.'[19] The normalization of surveillance remains an important issue that needs further investigation is 'mission creep'.

Modes of Surveillance as Practice

One of the most critical issues related to the political economy of information today is that of surveillance of both individuals and populations carried out by agencies in the private and public sectors. Just as individual consumers and their consumption habits are of economic interest to an array of companies, governments too are keenly interested to 'sort out' individuals and communities that are perceived to pose a risk to national security, often through surreptitious means, and register entire populations in an effort to streamline welfare, nationality, and citizenship. Biometric data based on profiling, iris scans, facial profiling, DNA, and fingerprinting are now routinely produced, often on the basis of PPPs, and the resulting data sets are both commercially and politically

---

[19] Wood, D.M. 2013. 'What Is Global Surveillance? Towards a Relational Political Economy of the Global Surveillant Assemblage', *GeoForum* 49: 323.

significant. While threats to national security have been amplified in the aftermath of wars being fought by Western countries in the Middle East, in particular, 9/11 and terrorist attacks such as in Mumbai, the failure of the West to be able to create lasting frameworks for democracy in Iraq and Afghanistan is a significant reason for contemporary investments in surveillance. In other words, arguably, the rationale to protect the nation and 'nationals' stems from the failure of foreign policy and its resulting instabilities. While the monitoring the data that individuals generate via Facebook, Skype, mobile phone calls, and emails is relatively easy to effect as the Internet service providers (ISPs) and mobile phone companies are now expected by law to preserve all data traffic for the period of a year or more and furnish encryption codes to governments—Blackberry had to hand over its encryption codes to the Indian government in 2012—the next phase, if we are to go by contemporary manifestations of surveillance, is based on it becoming the responsibility of all citizens.

The US Department of Homeland Security's Cell All System initiative is a public–private venture related to the creation of a 'personal environmental threat detector system consisting of multiple sensors which are miniaturized into a device and applied to an individual's cell phone.'[20] It will 'allow its users to continuously test the surrounding environment to harmful substances.'[21] In the case of the USA, this is the logical conclusion of steps such as the Warning, Alert and Response Network Act that enables commercial mobile service providers to transmit emergency alerts to their subscribers. As one of the private companies involved in standardizing algorithm and data management processes for alert and warning systems Telecommunication Systems, Inc. (TCS) proudly proclaims: 'As a thought leader and current member of the Communications Security, Reliability and Interoperability Council (CSRIC) committee, TCS is dedicated to developing innovative solutions to ensure public alerts are available when and where required.'[22] There are a number of other companies involved in

---

[20] US Department of Homeland Security, 'Privacy Impact Assessment for the Cell All Demonstrations'.

[21] US Department of Homeland Security, 'Privacy Impact Assessment for the Cell All Demonstrations'.

[22] 'TeleCommunication Systems Selected by U.S. Department of Homeland Security to Research Improvements in Geo-targeting Accuracy for Wireless Emergency Alerts'. 2015, *ComTech*, 13 November, available at http://www.

such projects in the USA, including Accenture that leads the $10-billion Smart Border Alliance and Ericsson that has been contracted to implement Strategic Border Initiative. Such extraordinary partnerships between the State and citizen are possible precisely because there is a widespread inter-sectoral perception of the randomness of terrorist attacks, best illustrated by the fact that the thousands of CCTV cameras installed in London did not make a difference in the prevention of the attack on the underground system, although they were certainly useful in identifying the bombers after the event.

It is quite extraordinary to come to grips with the types of prosumer behaviour of individuals that are generating data which are of interest to private companies and governments and to the privatization of 'watching' behaviours, self-watch, and neighbourhood watch. It is clear that across the world, privacy laws have been diluted, left in abeyance, or countenanced by reasons of state security, resulting in the fact that the data we generate online through whatever means is no longer 'private' but can be freely monitored, often by companies that provide the infrastructure of digital natives throughout the world. In fact, the Snowden revelations implicated all major Internet, computer, and telecommunications companies for proxy surveillance activities on behalf of the National Security Agency (NSA).

Historicizing Surveillance in India

The best-known history of surveillance in India is Christopher Bayly's *Empire & Information: Intelligence Gathering and Social Communication in India, 1780–1870.*[23] He has argued that the British built on and expanded the surveillance regimes established by their Hindu and Mughal predecessors by investing in running spies, political secretaries, and native informants. However, the sheer catholicity of the Indian oecumene was difficult to police and this led to the Indian Mutiny in 1857 that heralded

---

comtechtel.com/news-releases/news-release-details/telecommunication-systems-selected-us-department-homeland; accessed on 12 March 2019.

[23] Bayly, C.A. 1999. *Empire & Information: Intelligence Gathering and Social Communication in India, 1780–1870*, Cambridge Studies in Indian History and Society. New Delhi: Cambridge University Press.

the end of British rule in India. In other words, it was the 'autonomous networks of social communication within Indian society' that proved impervious to State surveillance and it was these 'affective communities of religion, belief, kinship, pilgrimage, literary or linguistic sensibility and style of political debate' that the State surveillance failed to co-opt.[24] Colonial anxieties about the 'Other' included the general population, but also very specific groups such as criminals and *badmaashes* (bad characters). Radhika Singha, on police surveillance in colonial India, makes the point that: 'A list of *badmaashes* (trouble makers) kept at every police station was a long-standing practice of colonial "preventive policing". Formalized in a 1793 regulation which required a money bond and sureties from those suspected of being vagrants, robbers, or "disorderly and ill-disposed" persons.'[25]

There are corollaries between surveillance then and now. Surveillance was and is about maintaining order in society and curbing the disorder that has been brought about by a variety of badmaashes. Today, the dominant badmaash is the terrorist who, unlike criminals in the previous eras who were located in space and place, is anonymous and part of the global population. Collecting information then and now on suspect populations is a means of control. Just as the British tried to make knowable the vast unknowable knowledge of India, today, surveillance is all about understanding the patterns of speech and conversation online. Giuliani, in a PhD thesis on surveillance in colonial Bengal, enumerates the extent and nature of surveillance and the relationship between knowledge and power:

> In a range of institutional contexts, information was administrative power. Historiography; the 'enumerative technology' of the census; the official efforts of the Company to master Indian languages, and the survey projects that sought to map, bind, and describe the physical, natural, and sociological features of India, all transformed knowledge into power. They produced information that was used to assess and collect taxes, to administer the law, and to identify, classify and group people into useful taxonomies such as 'tribe', 'village', and 'caste', such that they could be located and controlled.[26]

[24] Bayly, *Empire and Information*, p. 366.

[25] Singha, R. 2014. 'Punished by Surveillance: Policing "Dangerousness" in Colonial India, 1872–1918', *Modern Asian Studies* 49(2): 243.

[26] Giuliani, E.M. 2012. 'Policing Knowledge: Surveillance in Colonial India, 1861–1913', Thesis submitted for the Doctor of Philosophy, School of History, Religion, Philosophy and Classics. Australia: University of Queensland, p. 37.

The British were also involved in collecting information on their subjects in India through the census. By 1881, they had already begun collecting basic information that included not just names but also details on caste, religion, current residence, literacy, place of birth, and so on. Cohn has referred to this as the 'enumerative modality',[27] an initiative that foreshadows the UID project that involves the digitization of 'knowledge' on all residents of India. Post the 1857 revolt, the British created two legislations that enabled the surveillance of both 'voice' and 'text'. These were the Telegraph Act, 1885, specifically Section 5, and the Indian Post Office Act, 1898, Section 26. These two Acts allowed the colonial government to intercept messages and articles that were posted. While the Telegraph Act was amended in 1981, the text has not changed much and the State continues to use the phrase 'public emergency' to justify surveillance. In fact, there has been an attempt to normalize the State's digital surveillance through measures such as the CMS by suggesting that this is merely the 'automation' of existing systems of interception and monitoring.[28] Indeed, the powers invested in the CMS are expansive; the CMS is neither backed up by legislation nor is answerable to Parliament. As Kaul points out, the CMS can tap into many feeds and:

> monitoring and interception schemes across India. These include the Crime and Criminal Tracking Network and Systems (CCTNS), Lawful Intercept and Monitoring Program (LIM), Telephone Call Interception System (TCIS) and the Internet Monitoring System (IMS). The various department/ agencies that will have access to all this gathered data, through CMS, include the Central Bureau of Investigation (CBI), Defence Intelligence Agency (DIA), Department of Revenue Intelligence (DRI), Enforcement Directorate, Intelligence Bureau, Narcotics Control Bureau, National Intelligence Agency, Central Board of Direct Taxes, Ministry of Home Affairs, the Military Agencies of Assam and Jammu & Kashmir, and the Research and Analysis Wing (RAW).[29]

[27] Cohn, B. 1996. *Colonialism and Its Forms of Knowledge: The British in India*. Princeton: Princeton University Press, p. 8.

[28] Acharya, B. 2013. 'Turning India into a Surveillance State I', *The Hoot*, 18 November, available at http://www.thehoot.org/web/Turning-India-into-a-surveillance-state-I/7149-1-1-12-true.html; accessed on 25 May 2018.

[29] Kaul, M. 2014. 'Ensuring Privacy in a Regime of Surveillance', *Seminar* 655: 22.

While the CMS will monitor traffic over mobile phones, landlines, and the Internet, other surveillance initiatives include: the Network Traffic Analysis Initiative (NETRA), 2014, that will monitor Internet traffic, Skype calls, and so on; the National Intelligence Grid (NATGRID) that will use data analytics to make sense of big data from 21 government databases that belong to various government ministries; the Crime and Criminal Tracking Network System (CCTNS) that facilitates the sharing of data across 21,000 police stations in India; National Counter-Terrorism Centre (NCTC); the Indian Computer Emergency Response Team (CERT); the Lawful Intercept Monitoring Systems; and the Telecom Enforcement and Resource Monitoring Cells; among others.[30]

Surveillance Partners: India and the USA

India's investment in surveillance has been precipitated by international events, such as 9/11, and a variety of domestic acts of terrorism, including the storming of Parliament in 2001 and the Mumbai terrorist attacks in 2008. These investments are also a reflection of the new-found consanguinities between the USA and India in areas of defence and the war against terrorism. An article in *Jane's Defence Weekly* highlights the partnership between an Indian company, Dynamatic Technologies, and AeroVironment, a US developer of unmanned aerial vehicles (UAVs), which was officially inaugurated on 17 February 2015 as part of the — US–India Defense Technology and Trade Initiative (DTTI).[31] This is one project that is part of a suite of 22 proposals for joint military initiatives that are currently being negotiated—a process that has accelerated after

---

[30] See *India's Surveillance State*, World Wide Web Foundation, New Delhi, n.d., pp. 1–60, available at http://sflc.in/wp-content/uploads/2014/09/SFLC-FINAL-SURVEILLANCE-REPORT.pdf; accessed on 21 April 2018; and Xynou, M. 2014. 'Big Democracy, Big Surveillance: India's Surveillance State', *Open Security*, 10 February, available at https://www.opendemocracy.net/opensecurity/maria-xynou/big-democracy-big-surveillance-indias-surveillance-state; accessed on 25 April 2018.

[31] Grevatt, J. 2015. 'Indian and US Firms Inaugurate Joint UAV Development Capacity', *ISH Jane's 360*, 17 February, available at http://www.janes.com/article/49020/indian-and-us-firms-inaugurate-joint-uav-development-facility; accessed on 22 April 2018.

President Obama's visit to India in January 2015 as chief guest at its Republic Day celebrations. As part of this deal, India will export drones and spy equipment to the USA, while the US-based AeroVironment Inc.'s Raven drone and Lockheed Martin Corp.'s 'Roll On–Roll Off' kits, which turn jumbo transport jets into surveillance aircraft, will in turn be manufactured in India.[32] In 2017, the Trump administration approved a $2-billion sale of 22 sophisticated drones to the Indian government, the first to a non-North Atlantic Treaty Organization (NATO) partner.[33] All of this confirms that the USA's FDI in Indian defence has been steadily increasing. While DTTI is one among many joint US–India partnerships, another important one is the Indo-US Strategic Dialogue that began in 2009 on issues of mutual interest, including energy and climate change; education and development; economy, trade, and agriculture; science and technology; health and innovation, and has been elevated to Strategic and Commercial Dialogue in 2015. It also includes support for counter-insurgency training given that India is involved in conflicts on its borders with Pakistan and China and internally in Kashmir, north-east India, and the Maoist movements in central and east India.[34]

Sumit Kumar has also highlighted the extent of US–India security-related cooperation:

The India–US Counter Terrorism Cooperation Initiative seeks to further enhance the cooperation between two countries in Counter Terrorism as an important element of their bilateral strategic partnership. The initiative, inter alia, provides for strengthening capabilities to effectively combat

---

[32] Bipendra, N.C. 2015. 'India said to Plan Army Drone Exports to U.S. in Role Reversal', *Bloomburg Business*, 25 March, available at http://www.bloomberg. com/news/articles/2015-03-24/india-said-to-plan-army-drone-exports-to-u-s-in-role-reversal; accessed on 22 April 2018.

[33] Townshend, A. 2017. 'Prosperity through Partnership: US and India Talk Up Economic and Strategic Ties', *The United States Study Centre*, University of Sydney, 27 June, available at https://www.ussc.edu.au/analysis/prosperity-through-partnership-us-and-india-talk-up-economic-and-strategicties; accessed on 21 April 2018.

[34] For Increased US Investments for Counter-Insurgency Support in India, see the Plea by Mullick, H.A.H. 2013. 'The Naxalite Rebellions', *The American Interest* 9(1), available at http://www.the-american-interest.com/2013/08/11/the-naxalite-rebellions/; accessed on 21 April 2018.

terrorism; promotion of exchanges regarding modernization of techniques; sharing of best practices on issues of mutual interest; development of investigative skills; promotion of cooperation between forensic science laboratories; establishment of procedures to provide mutual investigative assistance; enhancing capabilities to act against money laundering, counterfeit currency and financing of terrorism; exchanging best practices on mass transit and rail security; increasing exchanges between Coast Guards and Navy on maritime security; exchanging experience and expertise on port and border security; enhancing liaison and training between specialist Counter Terrorism Units including National Security Guard with their US counter parts.[35]

Additionally, in 2011, an MoU was signed between the Indian and the US CERTs to exchange information on cybersecurity, system vulnerabilities, and joint technology development.[36] The two countries have been involved in a Cyber Security Forum and have agreed to a CERT, and India has participated in a cyber war game hosted by the Department of Homeland Security.[37]

While Edward Snowden's revelations included evidence of the NSA spying on Indian politicians, including those belonging to the incumbent party, it would seem that this issue is no longer a priority for the Indian government. In fact, the previous external affairs minister, Salman Khurshid, actually defended the US government's surveillance programme when he said: 'it is not actually snooping' and 'This is not scrutiny and access to actual messages. It is only computer analysis of patterns of calls and emails that are being sent. It is not actually snooping on specifically on content of anybody's message or conversation [sic].'[38] This after India was the sixth most tracked country under the NSA's programme that

[35] Kumar, S. 2012. 'India–US Relationships: From Estranged Democracies to Strategic Partnership', *Research Today* 1(1): 57–8.

[36] Kumar, M. 2011. 'India–US Sign Cyber Shield Deal', *The Hacker News*, 20 July, available at http://thehackernews.com/2011/07/india-us-sign-cyber-shield-deal.html; accessed on 20 April 2018.

[37] McArdle, J. and M. Cheetham. 2014. 'Indo-US Cyber Security Cooperation', *Seminar* 655: 28–31.

[38] *The Hindu*. 2013. 'It Is Not Actually Snooping: Khurshid on US surveillance', 2 July, available at http://www.thehindu.com/news/national/it-is-not-actually-snooping-khurshid-on-us-surveillance/article4873351.ece.

yielded 6.3 billion pieces of intelligence![39] There are those who have argued that official Indian reticence to criticize the US government was due to the fact that India itself had launched its own CMS to monitor phone calls and online traffic: 'In siding with the U.S., India is sending a signal to any potential Edward Snowdens employed within in its own security apparatus [sic]'.[40] The fact remains that the Indian government refused the asylum application from Snowden who had revealed the extent of PRISM[41]-related spying in India. One of the outcomes of these revelations was a call for public investments in domestic servers. While the Supreme Court of India did hear a public interest litigation (PIL) that called for accountability from companies in India that had disclosed information to the NSA, it was of the opinion that it did not have the power to take legal action against the US government or its companies that were incorporated in India.[42]

Surveillance in the Absence of Privacy Laws

There have been major investments in the surveillance of entire cities. Surat, Gujarat, became the first city to be equipped with a network of CCTV cameras and there are plans to roll this out in other cities in Gujarat. In 2013, as mentioned earlier, the ASSOCHAM and PwC released a document in

[39] Bajoria, J. 2014. 'India's Snooping and Snowden', The Wall Street Journal, 5 June, available at http://blogs.wsj.com/indiarealtime/2014/06/05/indias-snooping-and-snowden/; accessed on 21 March 2018.

[40] Muzaffar, M. 2013. 'Why India Is Taking the US's Side in the Snowden Scandal', The New Republic, 4 July 2013, available at https://newrepublic.com/article/113764/why-india-taking-uss-side-snowden-scandal#.

[41] PRISM is the code name for the surveillance operation carried out by the NSA in the USA which allows the collection of data directly from the servers of large digital corporates including Google, Microsoft, Facebook, and others (See Lee, T.B. 2013. 'Here's Everything We Know about PRISM to Date', The Washington Post, 12 June, available at https://www.washingtonpost.com/news/wonk/wp/2013/06/12/heres-everything-we-know-about-prism-to-date/?utm_term=.ca2bc3b967dd; accessed on 12 March 2019.)

[42] Hickok, E. 2013. 'What India Can Learn from the Snowden Revelations', CIS, 23 October, available at http://cis-india.org/internet-governance/blog/yahoo-october-23-2013-what-india-can-learn-from-snowden-revelations; accessed on 21 April 2018.

which the accent was on Ministry of Home Affairs and state government
subsidies and PPPs to build capacities in state surveillance through
investments in surveillance equipment, network connectivity, data centres
(servers, storage, applications), command viewing centres, collaborative
monitoring, and change management and capacity building.[43] There are
at least 76 private companies involved in equipping this sector, mainly the
supply of CCTV cameras and biometric technology. Maria Xynou, from
The Centre for Internet & Society (CIS), highlights the threat from large-
scale surveillance in the absence of strong privacy laws:

> This is extremely concerning because India lacks privacy legislation which
> could safeguard individuals from potential abuse. The fact that India has not
> enacted a privacy law ultimately means that individuals are not informed when
> their data is collected, who has access to it, whether it is being processed, shared,
> disclosed and/or retained. Furthermore, the absence of privacy legislation in
> India also means that law enforcement agencies are not held liable and this has
> an impact on accountability and transparency, as it is not possible to determine
> whether surveillance is effective or not. In other words, there are currently
> absolutely no safeguards for the individual in India and simultaneously, the
> rapidly expanding surveillance industry poses major threats to human rights.[44]

In the absence of explicit privacy laws guaranteed by the Constitution,
this right has been read into the right to freedom (Article 19) and the
right to life and personal liberty (Article 21). The promulgation of the IT
Act, 2000, and in particular IT Rules, 2011, under this Act, attempts to
provide privacy rights for individuals (Section 43A) in the context of an
accelerated collection of personal information by a variety of corporates.
These Rules require corporates to actively explore informed consent and
to abide by clear limitations as to the use of personal data. These rules
and statutes related to surveillance include: IT *(Due Diligence Observed
by Intermediaries Guidelines) Rules, 2011*; IT *(Reasonable Practices and
Procedures and Sensitive Personal Information) Rules, 2011*; and IT
(Guidelines for Cyber Cafe) Rule, 2011. However, this Act applies strictly
to the private business and corporate sector and does not apply to the
government that, arguably, is privy to the largest database on personal

---

[43] ASSOCHAM and PwC, *Safe Cities*.
[44] Xynou. 'The Surveillance Industry in India'.

information in India. One of the most controversial aspects of the IT
Act, 2000, has been Section 66A that penalized the sending of offensive
messages online. This section has been invoked more often than not for
political reasons and has led to numerous arbitrary arrests of individuals
for posts that were construed to be offensive to politicians. After a number
of petitions against the unconstitutionality of Section 66A, the Supreme
Court decided in favour of removing it in March 2015. While this was
an important victory for free speech, there are a range of clauses between
65 and 71 in these IT Rules that can be invoked by the government
to prosecute people for national security reasons. While successive
communication legislations in India prohibit the interception of personal
information, these legislations also include ambiguous language that
allows for the abeyance of privacy, in the light of security concerns,
public safety, and public emergency. In the light of concerns related to
cybersecurity, the Internet Services Licence Agreement gives the State the
power to monitor ISP traffic and 'Clause 34.12 and 34.13 requires the
ISP to make available a list of all subscribers to it's services on a password
protected website for easy access by Government authorities.'[45]

The State's involvement in curtailing the freedom of expression on
the Internet and privacy has been documented in the report made by
the Internet Democracy Project.[46] Privacy in India is complicated by the
Right to Information (RTI) Act, 2005. While this Act gives all citizens the
right to demand a transparent and accountable government, the Act does
exclude the police and security agencies and there have been a number
of public, mediated discussions on the unwillingness of judges from the
Supreme Court and major political parties to voluntarily come under
the purview of this Act. In fact, there have been attempts to amend this
Act in the light of such protestations. While privacy-related disclosures
are only applicable to the business and corporate sector and RTI-related

[45] *Privacy in India: Country Report*, CIS, October 2011, available at http://
cis-india.org/internet-governance/country-report.pdf; accessed on 21 March
2018.

[46] 'India's Universal Periodic Review: Third Cycle', Stakeholder Report
by the Internet Democracy Project, New Delhi, 2016, available at https://
internetdemocracy.in/wp-content/uploads/2016/10/Internet-Democracy-
Project-submission-for-Indias-UPR-Review-3rd-cycle_6thOct.pdf; accessed on
11 November 2017.

disclosures to the government bureaucracy, key state-related institutions involved in surveillance and data mining are not accountable and this remains a serious lapse.[47]

The UID (Aadhaar) Scheme

The creation of an authentic, reliable, and verifiable ID card for all citizens has attracted investments in both totalitarian and democratic countries. The UK government had, for example, experimented with ID cards in 1919 and 1939. The fixing of 'identity' has invited public investments, in countries keen to extend their authoritarian control over the citizens and in those that have extensive welfare entitlements for their citizens. A good example of the former is ID-based racial profiling in apartheid South Africa, which was linked to the system of homelands in which different races lived in prescribed areas and in which social mobility was severely restricted, especially for the black and coloured populations. In recent years, and in the context of austerity drives, there has been a renewed call for an ID system that prevents or at least lessens welfare fraud, acts as an immigration control and anti-terrorism measure, and as a one-stop measure of identification that can strengthen citizen–government relationships and exchanges. Today, a variety of digital techniques, from fingerprinting, to iris recognition, to DNA, have become the basis for biometric-based projects; and the scalability of such projects, in particular, has made it an attractive proposition for governments throughout the world. From the perspective of government, an ID scheme does have positive benefits: for instance, the Bhoomi land registration project in India that was based on biometrics led to the computerization of millions of land records in the southern state of Karnataka and, in the process, helped cut out the 'middleman's' role in controlling access to and the provisioning of land records—a process referred to as the 'disintermediation effect'.[48] However,

---

[47] Kaul, M. 2013. 'India: Right to Information and Privacy "Two Sides of the Same Coin"', *Index on Censorship*, 25 September, available at: https://www.indexoncensorship.org/2013/09/indias-right-information-act-colliding-privacy/; accessed on 10 April 2017.

[48] See Thomas, P.N. 2009. '*Bhoomi*, Gyan Ganga, e-Governance and the Right to Information: ICTs and Development in India', *Telematics and Informatics* 26(1): 20–31.

critics of such schemes point out the drawbacks of these schemes, especially if they are based on PPPs, namely, mission creep, and problems with the process of biometric identification that can result in the manufacture of 'truths' related to a person based on inaccurate data.

The complexity of identification is best illustrated by the UK government's initial support for a national entitlement card and its eventual jettisoning in the light of unresolved issues with the identification process. In 2002, the Labour government decided in favour of an ID card which led to the ID Cards Act, 2006. Fifty pieces of information were incorporated into this card that was the backbone of UK's National Identity Scheme (NIS). The Labour government, rather unproblematically, accepted the reliability of biometric technology and portrayed its benefits as follows:

> According to the Identity and Passport Service, the NIDS entails numerous and wide ranging benefits, both for society as a whole and for the individual. Among these benefits, the NIDS is expected to help protect cardholders against identity theft and fraud, provide a secure way of making financial transactions, and help to confirm eligibility for public services. The benefits to society include helping to prevent organized crime and terrorism, combating illegal working, reducing illegal immigration into the United Kingdom, and allowing police to more quickly identify suspects and people they arrest.[49]

Basically, the government downplayed the weaknesses that were evident in the actual process of data collection. Neyland has pointed out that the success rate of collecting verifiable information through facial recognition systems is around 73 per cent, along with errors related to fingerprinting of those with illnesses such as Parkinson's and those with hand injuries, and 5.6 per cent non-recognition of iris scans.[50]

A number of organizations, including a research team at the London School of Economics and an activist group, NO21D, opposed the NIS in terms of its rising costs, unreliability, and privacy-related concerns and consequences; also, the scheme was contrary to the principles laid out in the European Convention on Human Rights. Labour government

---

[49] Morris, J. 2008. 'Big Success or "Big Brother"?: Great Britain's National Identification Scheme before the European Court of Human Rights', *Georgia Journal of International and Comparative Law* 36: 467.

[50] Neyland, D. 2009. 'Who's Who? The Biometric Future and the Politics of Identity', *European Journal of Criminology* 6(2): 144.

spokespersons were not able to deal with the incongruities and inconsistences of the scheme, including the observations by NO21D that benefit fraud was based on under-reporting taxable income rather than ID fraud. In the light of a clear antagonism against the unpopular scheme, the Conservative government abandoned this £4.5 billion scheme in 2010—their first legislation in Parliament—destroyed the national register database that underlaid this scheme, and had to settle with the five companies awarded a stake in the scheme, including Computer Science Corporation (CSC), Electronic Data Systems (EDS, which had a previous history of botched IT projects at the Department of Inland Revenue and the Child Support Agency, UK), Fujitsu, IBM, and Thales SA, the French defence contractor.[51]

While issues related to identification were a priority for India's colonial rulers (fingerprinting evolved in India as an administrative tool for classifying colonized populations), the issue of identification has especially become focused in the aftermath of moral panics related to illegal immigrants, welfare fraud, and particularly the threat of terrorism that was precipitated by the Mumbai terrorist attacks in 2006. In fact, the plan for an identity scheme had been included under the National e-Governance Plan adopted by the Indian government as early as 2006—a Multipurpose National Identity Card—although it was shelved. The Indian government's embrace of biometrics can be seen as the logical extension of the developmental state given its involvement in the nationwide extension of welfare support, from the distribution of essential grains and oils through its public distribution system (PDS) to its distribution of cooking gas and disbursement of guaranteed employment on public infrastructure projects for its poor. It is also, at the same time, a clear signal of the government's conviction in technology-based solutions. The fact that Nandan Nilekani, ex-chief executive officer (CEO) of Infosys, one of India's largest software companies, was appointed as the CEO of the Aadhaar project reflects the faith placed in India's technocrats. Arguably, it is also a reflection of the 'biometric state' in which the need to keep a close watch over its populations has become a major reason for the existence/survival of the State.

---

[51] See Martin, A.K. and K.P. Donovan. 2014. 'New Surveillance Technologies and Their Publics: A Case of Biometrics', *Public Understanding of Science* 24(7): 842–57.

The Unique Identification Authority of India (UIDAI) was established in 2009 under the aegis of the Planning Commission of India (a body that has since then been disestablished).[52] The data collected under the UID is supposed to complement data collected by the National Population Register (NPR), a scheme launched in 2010 to collect information from all residents in India aged five and above that will be included in a National Register of Citizens. While the collection of data by the NPR does have a political agenda—curbing and monitoring illegal immigrants in India, especially from Bangladesh—the UID is supposed to be neutral and is often described in the language of 'efficiency' and 'transparency'. However, it is unclear as to what extent the UID card will be used by the private sector to vet access to resources and monitor transactions. This is an issue that has not been adequately researched in India. While the UID is voluntary and not mandatory, all residents in India are mandated to provide entries under the NPR. Also, the NPR is a countrywide project, whereas the UID is currently being rolled out in 17 states and 2 union territories. Unlike the NPR which has a legal basis through the Citizenship Act, 1955, and Citizenship Rules, 2003, the UID project has not been backed up in legislation and there is continuing debate on the legality of the enterprise in the absence of such legislation. While the official literature on the UID indicates that the disclosure of information by individuals is entirely voluntary, the architecture of this project and its ecosystem encourages enrolment, is intentionally extensive, and involves integration of its Central Identities Data Repository (CIDR) with existing databases at central and state levels currently run by public and other bodies, thus enabling the government to access vast amounts of personal data that can be used to profile suspect individuals—both those who 'rort' the system and who are suspected for their anti-national behaviour. One of the issues with the UID is that it currently does not include robust data and privacy protection measures. An equally worrying trend is public investments in a variety of large national databases, in which the line between data for functional use and for surveillance has got blurred. Das, Mitra, and Bagchi have highlighted another database initiative that is locked into the UID:

> The 'national intelligence grid' (NATGRID) project is a similar endeavour envisaged by the government [that] will interlink 21 categories of databases

52 See https://uidai.gov.in/; accessed on 20 March 2018.

(railway and air travel, income tax, phone calls, bank account details, credit card transactions, visa and immigration records, property records, driving licence etc) for real-time monitoring of all residents in the country, and will logically use UID number as the inter database linkage.[53]

The UID project has received a lot of attention in the media because it is the largest-ever effort at creating a comprehensive national data register that is based on linkages with a battery of personal data nodes, including passport, driver's licence, bank accounts, and so on. In other words, its promise of interoperability is an intriguing prospect for governments interested in rolling out such a scheme. As Sarkar has observed: 'Aadhaar attempts to map the population of a given territory completely (mapping residents, not citizens) and to render the population as a rational legible space, without any fakes and duplicates. It can further accommodate various vectors—cartographic, ethnographic, demographic—each of which can be added to information on a unique body.'[54] It is also an attempt to transcend earlier ID schemes that were place-based. The UID scheme is built to assist both the stationary and mobile Indian, whose ID and entitlements can be validated at any point throughout the country. Its detractors have argued that its comprehensiveness can have consequences for privacy. The government has also invested in the infrastructure, technology, and services required to collect data on every Indian citizen, outsourced its software requirements, and provided financial services, in particular an opportunity to create larger markets. Companies involved in providing these services include Mindtree and Intelenet Global Services. The scale of the outsourcing has itself generated a variety of anxieties given that private firms have access to personal data.

While the intent of the UID is clear enough, its implications for the ordinary citizen are less than clear. The State essentially sees the UID as a technological solution that would enable every Indian citizen to be invested with a UID code and number that can be invoked, policed, and managed. If, as is the case, more and more data on individuals continue to be added, then there is the likelihood that the UID card will enable the State to

---

[53] Das, J.K., S. Maitra, and D. Bagchi. 2011. 'Unique Identification Number: The New Identity Paradigm', *Global Management Journal* 5(1–2): 13.

[54] Sarkar, S. 2010. 'The Unique Identity (UID) Project, Biometrics and Re-Imagining Governance in India', *Oxford Development Studies* 42(4): 525.

exercise its function as a Panopticon. Creating a singular identity, valid for all time, on the basis of biometric information does not account for the fact that errors can result in the construction of fictional identities—one reason for the UK government to shelve the scheme. Arguably, such collections of discrete, personal data facilitates the selective sense making of bodies that can be controlled and managed by governments. The general threat of terrorism has provided the means for the State to embark on this scheme, thus assuaging real and imagined fears faced by ordinary citizens. And yet, the fact that the UID's 'sorting' and interpellation involves the rendering of both functional, day-to-day entitlements and the suspect welfare cheat and terrorist highlights the power and reach of such centralized, technology-based databases. In the words of Jacobsen:

> The combination of surveillance and inclusion in the UID assemblage produces several objects of knowledge: the financially excluded, the fraudulent and the unidentified 'poor and marginalized' who are the principle 'moving targets' …: of the system. Through establishing biometric-based IDs as a recognized means of identification, the UID assemblage thereby incorporates the possibilities of both empowerment and confinement, recognition and disapproval.[55]

Arguably, the implementation of large-scale State surveillance through initiatives such as the UID results in the dispossession of the personal and the private, processes that complement the dispossession of material assets such as land. There is as yet no sign of a double movement against surveillance, since opinion is divided on the merits of surveillance in the context of heightened animosities against minorities such as Muslims, who are seen as a threat against the Hindu nation. One of the fascinating dimensions of surveillance is the market as a willing accomplice in supporting, investing in, and legitimizing the State surveillance. Foster and McChesney, in an article on 'surveillance capitalism' in the USA, refer to the many ways in which corporate institutions in the USA cooperate with the US military and the State in surveillance initiatives, thus enhancing the reach of the surveillance State.[56] This is also the

---

[55] Jacobsen, E.K.U. 2012. 'Unique Identification: Inclusion and Surveillance in the Indian Biometric Assemblage', *Security Dialogue* 42(5): 468.

[56] Foster, J.B. and R.W. McChesney. 2014. 'Surveillance Capitalism: Monopoly-Finance Capital, the Military–Industrial Complex and the Digital Age', *Monthly Review* 66(3): 23–4.

case in India, although there is as yet no study that explores the extent to which corporates have become part of the surveillance State. When surveillance becomes a matter of patriotism and security, it becomes difficult for society to challenge this frame; and this is certainly the case in India where issues related to privacy have not become part of everyday conversation. In other words, State surveillance has not been countered effectively precisely because these are processes that are a manifestation of the power of the State to resist scrutiny and accountability. Abrams captures this aspect of the secret State that is ultimately difficult to know precisely because it has the power to resist an in-depth investigation of its workings: 'Any attempt to examine the politically institutionalized power at close quarters is, in short, liable to bring to light the fact that an integral element of such power is the quite straightforward ability to withhold information, deny observation and dictate the terms of knowledge.'[57]

While close to 600 million people have been registered under the Aadhaar scheme, enrolment has been problematic. Usha Ramanathan has observed: 'the UID is entirely based on outsourcing, with embarrassing errors showing up, where dogs, trees and leaves have been issued UID numbers, and over 800 people have been enrolled as "biometric exceptions" in one episode in Hyderabad when they have been later found not to exist.'[58] A recent issue related to Hindu God Hanuman being assigned a UID number became fraught and the basis for an investigation by the Hindu national BJP government.

The process of biometric identification follows three steps. It begins with *enrolment* during which personal data is captured, processed, and stored for operational purposes. *Biometric verification* is the process in which a claimed identity is verified against the available data and *biometric identification* occurs when biometric data is compared against a range of biometric records. In the case of the UID, these three processes have been complex—and the complexity is in some measure due to the extraordinary diversity of the population in India and issues related to recording that diversity. One of the more critical, academic writings on

[57] Abrams, P. 1988. 'Notes on the Difficulty of Studying the State', *Journal of Historical Sociology* 1(1): 62.

[58] Ramanathan, U. 2013.'A Tale of Two Turfs: NPR and UID', *The Statesman*, 10 October, available at http://www.thestatesman.com/news/10497-a-tale-of-two-turfs-npr-and-uid.html; accessed on 11 April 2018.

the UID is Rao and Greenleaf's article in *Surveillance & Society*.[59] While they acknowledge the laudability of the intention of the UID project that includes creating a UID that will help the poor get access to banks and welfare, their focus is on the fraught process of identification, in particular the relationship between the machine and body on the Indian street and the unintentional consequences of mission creep. Their fieldwork with the Delhi-based NGO called Homeless Service that helps with the enrolment of homeless people highlights the fact that the machine–body interface is conditioned by social norms. Women who are 'trained to lower their gazes or veil their faces' find it difficult to look into a camera and there were issues with people whose fingers were disfigured or scarred.[60] They also observe: 'UID is developing a social life on the streets. Here, coded technology meets vulnerable human bodies, and welfare institutions struggle to manage conflicts between institutions and poorly documented citizens during the roll-out of new welfare schemes.'[61]

The enrolment of the homeless has posed unique issues: for example, how does one issue a unique number in the absence of documents and a clear residential address? To deal with such cases, the UIDAI has come up with the idea of an 'introducer', someone who can vouch for a person who has no documents. While this is an ingenuous method, it nevertheless creates a culture of middlemen who have the power to 'enrol' those whose very survival depends on their access to entitlements—from rationed food to employment and shelter. It is written on the UIDAI website:

> much of the poor and under-privileged population lack identity documents and Aadhaar may be the first form of identification they will have access to. The UIDAI will ensure that its Know Your Resident (KYR) standards do not become a barrier for enrolling the poor and has accordingly developed an Introducer system for residents who lack documentation. Through this system, authorised individuals ('Introducers') who already have an Aadhaar, can introduce residents who do not have any identification documents, enabling them to receive their Aadhaar.[62]

---

[59] Rao, U. and G. Greenleaf. 2013. 'Subverting ID from Above and Below: The Uncertain Shaping of India's New Instrument of e-Governance', *Surveillance & Society* 11(3): 287–300.

[60] Rao and Greenleaf, 'Subverting ID from Above and Below', p. 294.

[61] Rao and Greenleaf, 'Subverting ID from Above and Below', p. 293.

[62] UIDIA, 'Why Aadhaar?', available at http://uidai.gov.in/why-aadhaar.html; accessed on 11 May 2018.

The Aadhaar site highlights the fact that the initiative is for all Indians, inclusive of transgenders and infants. This intent to enrol all Indians, irrespective of their class, caste, gender status, and location, can be considered progressive. However, the role of the middleman as midwife to the enrolment process does include risks given that the middleman has sole responsibility for verification. One of the more urgent issues from a privacy perspective is that of whether or not the UID's welfare and security focus will translate into actions against Indian citizens. Another is, recent judgements by the Supreme Court of India against the UIDAI sharing information with government agencies, recommending the de-linkage of UID from the country's welfare schemes, and not making the UID mandatory.[63] Furthermore, the present government's lukewarm response to the UID does suggest that the UIDAI will need to accommodate to changes, including the possibility that it is merged with the NPR. In the absence of any clear benefits for those who have been registered under the UID scheme, and growing confusion in government circles as to how to manage this multibillion-dollar project, an advertorial on an NGO portal indicates one of the likely winners: 'MasterCard Worldwide has collaborated with India's Unique Identification Authority to provide easy payment solutions for UID account holders with the objective of enabling financial inclusion of millions of people. The solution supports prepaid, debit and credit payment products and will empower "Aadhaar" account holders to move towards electronic transactions.'[64]

Code and the Welfare State: The Politics of e-Governance

While the Department of Electronics and Information Technology's (DeitY) 'Framework for Citizen Engagement in e-Governance' offers a guide to the principle and process of citizen engagement and of the need for a shared vision based on principles of citizen centricity, transparency,

---

[63] See Jayaram, M. 2014. 'India's Big Brother Project', *Boston Review*, 19 May, available at http://bostonreview.net/world/malavika-jayaram-india-unique-identification-biometrics; accessed on 11 January 2018.

[64] 'MasterCard Provides Payment Solutions for India's UID Card Holders', *OneWorld South Asia*, 10 December 2010, available at http://southasia.oneworld.net/archive/ictsfordevelopment/mastercard-provides-payment-solution-for-indias-uid-card-holders#.Vniuz6N--os; accessed on 10 January 2018.

and accountability,[65] the fundamental premise that the National e-Governance Plan (2006) is based on is that technology as intermediary, rather than human intermediaries, will mediate citizen's access to a range of government services across multiple sectors. The Plan highlights 26 'Topics' from agriculture to youth and sports and lists a variety of functions available: seeking access, checking, tracking applications, and so on. In India, e-governance has moved on from its initial accent on the computerization of government departments to the computerization of government services. The Plan includes 31 Mission Mode Projects (MMPs), one of which, 'e-Governance in Municipalities', states that:[66]

1. This has, at present, very limited or no computerization across urban local bodies (ULBs) in different states.
2. There is very limited or non-existent staff with IT know-how.
3. There is lack of standardization of process.
4. The processes are primarily operated in a manual mode.

In other words, this would suggest that e-governance in India began as a top-down project in which capacity building has not been the most important of priorities. While people in India, including the poorest, are accessing the digital particularly through mobile phones, given the relatively low Internet penetration and access to computing, it is more likely the case that intermediaries who have access to computers are involved in servicing the needs of the poor. Outside any government department of India, from courts to welfare offices, there are 'touts' involved in scribal activities that cater to the needs of illiterate populations. There are two kinds of intermediaries: (i) those who are 'corrupt' and 'act as a conduit of information between government officials and members of the public in the disbursement of a public benefit';[67] and (ii) those who are involved in

---

[65] DeitY. 2012. 'Framework for Citizen Engagement in e-Governance', April, pp. 1–28, available at http://www.archive.india.gov.in/allimpfrms/alldocs/16485.pdf; accessed on 11 April 2018.

[66] GoI. 2014. 'State MMPs', updated on 21 November, available at http://india.gov.in/e-governance/mission-mode-projects/state-mmps; accessed on 24 April 2018.

[67] Bose, G. and S. Gangopadhyay. 2009. 'Intermediation in Corruption Markets', *Indian Growth and Development Review* 2(1): 40.

doing a service and have either been appointed by the government or act on their own volition. The latter can also be corrupt given that they are involved in providing services to illiterate people whose survival depends on access to public goods and who are, more often than not, unaware of their rights as citizens of India. Both these types of intermediaries exist because of information and other asymmetries. So, arguably, e-governance in India has been built on an existing system that is founded on multiple divides—class, caste, and gender.

I have argued elsewhere that there is a belief that closing the digital divide will also result in the resolving of other divides in society. However, as it is plain to see in contemporary India, caste remains a primary identity in India and legislations and education have not been able to dent the power of caste as a way of ordering society. Caste blocs, for example, are key to understanding the electoral system in democratic India. In other words, there is little or no evidence that e-governance projects have led to equity in access or, for that matter, to making caste irrelevant. Rahul De, in an article on caste and e-governance related to three prominent e-government projects in India (Bhoomi, Gyandoot, and the Village Knowledge Centre), points out that the Bhoomi project, which is about the computerization of land records,

> followed the priorities of the dominant landed castes. It was designed to provide easy access to land records and to make efficient the land sale or transfer process. These benefited the land owning castes the most, as they were in the best position to use the easy availability of land records to obtain loans and also participate in land transactions. Dalit and lower castes work mainly as landless labor and as tenant farmers in the state and they had marginal use of the Bhoomi system … . Caste, an 'idiom of association,' manifested itself in everyday practices of access to, mobility around, sharing of knowledge about, and use of the kiosks. Caste affiliation and privilege played directly into the equation for extracting the new resources made available by a powerful technology. Everyday practices of the dominant castes facilitated, and were facilitated by, the easy appropriation of the technology.[68]

[68] De, R. 2009. 'Caste Structures and e-Governance in a Developing Country', in M.A. Wimmer, H.J. Scholl, M. Janssen, and T. Traunmuller (eds), *Electronic Government: 8th International Conference, EGOV 2009, Linz, Austria, August 31–September 3, 2009, Proceedings.* Berlin, Heidelberg: Springer, pp. 46, 51.

In other words, the State's embrace of code, its manifestations through thousands of e-government projects, and the justifications for its use are difficult to reconcile precisely because of a lack of political will to simultaneously deal with pre-existing divides in society in the context of these projects. There is romanticization about the capacity of e-governance to deliver when, in many cases, such projects merely accentuate the access divide.[69] One could argue that the centralized government that has characterized government in India has been able to use the technology-mediated language of transparency, accountability, efficiency, and access to protect and preserve its control over information flows. The achievement of transactional efficiencies has been a core concern for e-government at the expense of citizen involvement in participatory governance.[70] The private sector, which has been involved in and benefited from many of these public–private e-government prizes, has merely accentuated this myth through sponsoring annual e-government prizes, offering products for free, and, in the case of Microsoft, accentuating the commodification of resources that otherwise would have been shared.

> Supported by Microsoft, Jammu & Kashmir is rolling out cloud-based services to citizens. In what is an extremely interesting e-Governance rollout, the government has decided to share the existing e-Governance infrastructure and excess capacity (basically data center space) of the government of Madhya Pradesh. By transforming its data centers into a private cloud, the government of Madhya Pradesh can offer IT-as-a-service to any other state government, through what is expected to eventually become a pay-per-use model.[71]

Both code as surveillance and code in the provision of welfare point to the pitfalls of governance in India. While surveillance involves the

---

[69] See Martinez, J., K. Pfeffer, and T. van Dijk. 2011. 'E-Government Tools, Claimed Potentials/Unnamed Limitations: The Case of Kalyan–Dombivli', *Environment and Urbanization ASIA* 2(2): 223–34.

[70] Chatterji, T. 2017. 'Digital Urbanism in a Transitional Economy: A Review of India's e-Government Policy', *Journal of Asian Public Policy* 11(3): 334–49, available at https://doi.org/10.1080/17516234.2017.1332458.

[71] 'J&K to Take the Cloud Route to e-Governance, Share e-Governance Infrastructure with MP Government', *Microsoft Perspective*, n.d., available at http://www.microsoft.com/en-in/about/perspective/articles/a-e-governance. aspx; accessed on 20 January 2018.

deliberate exercise of State power to manage and control its populations, code as welfare is often described as the benign, paternal side of the Government of India, of a government investing in modernizing India and expanding access to all in a knowledge economy. William Mazzarella, in one of the better critiques of e-government in India, characterizes the entire enterprise as one that is built on what he terms the 'politics of immediation':

> E-governance is, it seems to me, one important avatar of a more general desire for what I am calling a politics of immediation—that is to say, a political practice that, in the name of immediacy and transparency, occludes the potentialities and contingencies embedded in the mediations that comprise and enable social life ... . Social practices of mediation, often initially quite contested, are formalized as mechanisms, externalized as technologies, and naturalized as social orders.[72]

The geopolitics of surveillance has become a subject for debate and discussion specifically in the aftermath of the Snowden revelations. However, State surveillance is only one side of the story given the extensive role played by social networks and search engines, such as Google, in the business of selling aggregated information on individuals to the market. Both methods of surveillance affect individuals, although this effect is not going to be uniform. Certain populations that are deemed a security risk will be of interest to the State, while those who have spending power will be of interest to search engines and social networks. While the State uses reasons of security and the private sector uses consumer sovereignty to embrace individual interactions with the Internet, arguably the disaggregated individual has become the focus for another digital enclosure that corresponds to other enclosures related to digital access and knowledge enclosures through measures such as intellectual property (IP). While issues related to surveillance are bound to increase in the coming years in the context of the globalization of the Internet and more people becoming digital natives, as the example of the Aadhaar scheme in India suggests, large-scale technological solutions are not foolproof and even the latest biometric technologies are unable to provide accurate

---

[72] Mazzarella, W. 2006. 'Internet X-Ray: e-Governance, Transparency and the Politics of Immediation in India', *Public Culture* 18(3): 476.

readings of individuals. There is, in an era of surveillance, a lot of room for error. As the issues with the pre-registration (enrolment) of homeless people in India demonstrate, incomplete data on individuals have already become a part of schemes such as Aadhaar. Moreover, people who have access to multiple homes can also become the owners of more than one ID card. Then, there is the issue of the copy, and fake IDs in the context of India are bound to mingle with the real ones. All this will have obvious consequences for governmentality. While recent efforts by the European Parliament to unbundle Google's search and advertising businesses might open the door for more competition in the search engine space,[73] that by itself will not make a dent on the privatization of personal data for which there is need for stronger privacy laws.

However, an equally important concern is the need to curb the power of the State and superstates to survey its own citizens and that of others. In the case of the Aadhaar scheme, there was an attempt to make the UIDAI a statutory body without the approval of Parliament. As Ramkumar has highlighted, the report of the Parliament's Standing Committee on Finance (SCoF), which examined the Bill to convert the UIDAI into a statutory authority, expressed strong reservations against this project:

> The report tears apart the faith placed on biometrics to prove the unique identity of individuals. It notes that 'the scheme is full of uncertainty in technology' and is built upon 'untested, unreliable technology'. It criticises the UIDAI for disregarding (a) the warnings of its Biometrics Standards Committee about high error rates in fingerprint collection; (b) the inability of Proof of Concept studies to promise low error rates when 1.2 billion persons are enrolled; and (c) the reservations within the government on 'the necessity of collection of IRIS image'.[74]

The report concludes that, given the limitations of biometrics, 'it is unlikely that the proposed objectives of the UID scheme could be achieved'.

---

[73] Oreskovic, A. 2014. 'Google Split Proposed in Bid to Level Playing Field', *Sun Herald*, 23 November, p. 23.

[74] Ramkumar, R. 2011. 'Aadhaar: Time to Disown the Idea', *The Hindu*, 16 December, available at http://www.thehindu.com/todays-paper/tp-opinion/aadhaar-time-to-disown-the-idea/article2719027.ece; accessed on 7 February 2018.

However and in spite of internal dissension, on March 2016, the lower house passed the Aadhaar (Targeted Delivery of Financial and Other Subsidies, Benefits and Services) Bill, hastening the legalization of mass surveillance. In the words of Sunil Abraham, CEO of CIS, Bengaluru: 'Aadhaar is mass surveillance technology. Unlike, targeted surveillance which is a good thing, and essential for national security and public order—mass surveillance undermines security. And while biometrics is appropriate for targeted surveillance by the state—it is wholly inappropriate for everyday transactions between the state and law abiding citizens.'[75] Despite government investments in cybersecurity, notably the Reserve Bank of India's cybersecurity framework for banks and the setting up of four sectoral CERTs specifically to protect power transmission and distribution, there are continuing issues with the security of large data sets on Indian citizens held with public bodies. These include the records of financial transactions of 20 million users held with the Goods and Services Tax Network (GSTN), information on 60 million taxpayers with the Income Tax Authority, passport details of 250 million citizens with Passport Seva, 1.19 billion UIDs with the UIDAI, information on 700 million online requests in 2017 alone for a variety of e-government services, along with a slew of state and central databanks with information on pension records, birth certificates, and so on.[76]

This chapter has highlighted the role of the State and State–private partnerships in the expansion of surveillance in India. Chapter 2 will explore the specific role played by the private sector in the development of the surveillance industry in India.

[75] Raman, A. 2017. 'Is Aadhaar a Breach of Privacy', *The Hindu*, 31 March, available at http://www.thehindu.com/opinion/op-ed/is-aadhaar-a-breach-of-privacy/article17745615.ece; accessed on 11 April 2018.

[76] See Singh, S. 2018. 'How Safe is Digital India? *The Economic Times*, 14 January, available at https://economictimes.indiatimes.com/news/economy/policy/how-safe-is-digital-india-indias-vast-data-pools-need-to-be-secured-with-tighter-de-risking-tools/articleshow/62489823.cms; accessed on 11 February 2018.

2

Leisure, Surveillance, and the Private Sector in India

This chapter focuses on surveillance and explores growth in the surveillance industry in India, its manifestation through PPPs, and the surveillance of 'affect' that is a core aspect of people surveillance practised by search engines and social networking sites.

At the core of India's neo-liberal developmental model is a turn towards PPPs in infrastructure development, industrial growth and manufacturing, and the knowledge economy. Public–private partnerships in e-governance have been the preferred model for over two decades, illustrated by flagship telecentre projects such as Gyandoot and e-Choupal, and the role of software companies such as Microsoft in providing much of the software for large-scale public projects. However, it is in the embrace of integrated knowledge-based infrastructure projects, such as the Smart Cities initiative unveiled by PM Modi in 2014, that we can see the extent of private sector involvement in the creation of cityscapes inclusive of the security paraphernalia that lies at the heart of such smart cities. Data surveillance has become a function of the State and is also key to business prospects. In fact it is often the case that these alliances of interests double up as business alliances. Luke Harding from *The Guardian* newspaper, in his book *The Snowden Files* provides startling evidence of the complicity of Silicon Valley's technology companies in NSA's PRISM project:

The first to provide PRISM material was Microsoft. The date was 11 September 2007. This was six years after 9/11. Next came Yahoo (March 2008) and Google (January 2009). Then Facebook (June 2009), PalTalk (December 2009), YouTube (September 2010), Skype (February 2011) and AOL (March 2011). For reasons unknown, Apple held out for five years. ... It joined in 2012. ... The top-secret PRISM program allows the US intelligence community to gain access to a large amount of digital information—emails, Facebook posts and instant messages.[1]

Some of the large social networking sites routinely pass on privately generated data on their networks to the Government of India.

Google and its social networking arm in India, Orkut, made a decision in 2007 to help security forces in India by providing them with information on Internet Protocol addresses and the ISP from which content seen as objectionable or against the national interest had been sent.[2] The fact that this included websites and blogs critical of the rabid, right-wing power-broker Bal Thackeray, murky Congress Party politics and deals, and the more than 150 members of parliament in the lower house who have criminal records reveals the extent to which the government is prepared to control access to and use of the Internet. In 2011, the Google 'Transparency Report'[3] revealed

> that during the period of June–January 2011, the Government of India had asked Google to get rid of as many as 358 items, in all, across its services like YouTube, Orkut, Google Earth, Google Maps, and Panoramio, Blogger and Picasa Web Albums. The report has Google stating of all the 'content removal requests' that Google received throughout the June–January 2011 period, it complied partially or fully to only 51 percent of these requests. ... The content removal requests that Google received fell under categories, like: National Security, Privacy and Security, Defamation, Impersonation, Pornography, Hate Speech, Government Criticism and Others.

Between July 2016 and December 2016, Google complied with 57 per cent of requests for data on accounts, a trend that has seen steady growth

---

[1] Harding, L. 2014. *The Snowden Files: The Inside Story of the World's Most Wanted Man.* London: Guardian Books and Faber & Faber Ltd, p. 198.

[2] 'Orkut's Tell-All Pact with Cops', *The Economic Times*, 1 May 2007, available at http://articles.economictimes.indiatimes.com/2007-05-01/news/28459689_1_orkut-ip-addresses-google-spokesperson; accessed on 12 August 2017.

[3] Shetty, A. 2011. 'Google Transparency Report Reveals Govt. Demanded Omission of 358 Items', *First Post*, 8 December, available at: https://www.firstpost.com/tech/news-analysis/google-transparency-report-reveals-govt-demanded-omission-of-358-items-3592301.html; accessed on 13 March 2019.

for over a decade.[4] During the first quarter of 2014, Google received 2,800 content takedown requests from the Indian government, indicating an exponential increase in such requests. Agrawal has observed: 'Most of the arrests for objectionable content on Facebook and Google … have taken place under sections 66(A) and 67 of the Information Technology Act, which relate to sending offensive messages through communication service and for publishing or transmitting of material containing sexually explicit act, etc. in electronic form.'[5] Bhargava, writing in *The Hindu*, has also reported that Facebook restricted access to 6,000 sites on the request of the Indian government during July–December 2014, the highest for any country.[6] Takedown requests, especially if they are a consequence of court orders, are often difficult to comply with, particularly for intermediary organizations, as they are merely involved in hosting content and not producing it. Given their relative powerlessness against the State, many of these intermediaries comply, resulting in extensive censorship of material deemed defamatory or obscene throughout the country. The spectre of blog posts leading to communal disharmony are often cited by the government as the reason for their action, although anti-Muslim sentiments are widespread, commonplace, and have become part of everyday discourses in India. The present government's strident Hindu nationalism contributes to a selective censoring of the Internet, which is supportive of taking down sites that 'slight' majority sensitivities, while allowing sites that bait minorities to exist and extend irredentist

---

[4] See IANS. 2017. 'Google Reports an All-Time High User Data Information Requests from Indian Govt., *ET Tech*, September 2017. Available at https://tech.economictimes.indiatimes.com/news/internet/google-reports-an-all-time-high-user-data-information-requests-from-indian-govt/60884339; accessed on 13 March 2019.

[5] Agrawal, V. 2015. 'Google Received 2,800 Content Takedown Requests from Indian in H1 2014', *The Times of India*, 11 February, available at http://timesofindia.indiatimes.com/tech/tech-news/Google-received-2800-content-takedown-requests-from-India-in-H1-2014/articleshow/46197554.cms?#write; accessed on 8 August 2017.

[6] Bhargava, Y. 2015. 'India Tops Facebook's Content Restriction Request List', *The Hindu*, 16 March, available at http://www.thehindu.com/sci-tech/technology/internet/india-tops-facebooks-content-restriction-request-list-for-second-time-in-a-row/article6999502.ece?homepage=true; accessed on 9 August 2017.

sentiments, especially in the case of the disputed territory of Kashmir. Melody Patry has commented on the consequences of Internet censorship:

> Google is not the only company dealing with a significant number of takedown requests. For small start-ups and internet service providers, a large number of takedown requests can encourage those afraid of penalties to over-comply, removing URLs that do not link to illegal content. A consequence of the IT Act and of the over-compliance would be the delegation of essential executive function to private parties like Google, Facebook or MouthShut.com to censor and restrict free speech of citizens or else face legal challenges over user content.[7]

This relationship between Internet-based companies and the State needs to be seen in the context of business opportunities arising out of the multibillion-dollar Digital India and Smart Cities projects launched by the BJP government in 2014. In the 2014–15 budget, the Indian government had allocated $1 billion to its Smart Cities project,[8] although its real costs are in the region of $1 trillion. The private sector is expected to mobilize resources, know-how, and technology in the creation of 22 smart cities by the year 2022. This project has already attracted interest from foreign and Indian companies. Some of the PPPs include: the CISCO and ILF Technologies Ltd's $90 billion infrastructure development project that includes an information and communications technology (ICT) master plan for four cities; Bloomberg Philanthropies and Indian Ministry of Urban Development's initiative aimed at citizen engagement in the building of these smart cities; and International Enterprise Singapore and the Infrastructure Corporation of the AP government in developing a master plan for the 7,325 sq. km new capital of AP.[9] This latter project is

---

[7] Patry, M. 2013. 'India: Digital Freedom under Threat? Online Censorship', *Index on Censorship*, 21 November, available at http://www.indexoncensorship. org/2013/11/india-online-report-freedom-expression-digital-freedom-1/; accessed on 11 August 2017.

[8] 'Smart Cities', available at http://www.makeinindia.com/article/-/v/ internet-of-things; accessed on 8 September 2017.

[9] Schenkel, S. 2015. 'Financing India's Smart Cities: The Case for Public–Private Partnerships', CogitASIA, Centre for Strategic & International Studies, 12 May, available at http://cogitasia.com/financing-indias-smart-cities-the-case-for-public-private-partnerships/; accessed on 17 September 2017.

the technocrat Chief Minister N. Chandrababu Naidu's latest project that involves the large-scale displacement of farmers. The project will involve the acquisition of 30,000 acres of arable land, some by enforcing eminent domain. This extensive multi-billion-dollar project is part of a much larger plan to develop smart cities throughout the state of AP: 'While two towns in East Godavari district—Kakinada and Rajahmundry—would be developed as Smart Cities, the other towns include Srikakulam, Vizianagaram, Guntur, Nellore, Prakasam, Anantapur and Kurnool. Also Kadapa town would be developed into [an] Industrial Smart City.'[10] Central to the concept of smart cities and its development is the role of major construction and other companies in the privatization and creation of 'safe' cities that are managed through wall-to-wall, 24-hour dataveillance. The 'command and control' centres that will monitor the functioning of public utilities, such as the sewer system, traffic, and people, are key to these cities.

One of the better examples of such a city is the Gujarat International Financial Tec-City (GIFT City) being built in Gujarat, PM Modi's home state. The accent is on building a post-modern city structured into a variety of enclaves: a financial district; a technology services district; and living areas with high-rise condominiums. It will have an SEZ, international education zone, integrated townships, an entertainment zone, hotels, a convention centre, an international techno park, Software Technology Parks of India (STPI) units, shopping malls, stock exchanges, and service units. As Ravindran has observed:

> The beating heart—or rather, robot brain—of Gift City is its 'Command and Control Centre', which keeps traffic moving smoothly and monitors every building through a network of CCTVs. In a country where more than 300 million people live without electricity, and twice as many don't have access to toilets, Gift City's towers sound like hypertrophic castles in the sky. But they are an essential part of the Indian government's urban vision, one that it wants to see replicated a hundred times across the country.[11]

---

[10] Mallikarjun, Y. 2014. 'Plan to Develop Fourteen Smart Cities', *The Hindu*, 5 September, available at http://www.thehindu.com/news/national/andhra-pradesh/plan-to-develop-14-smart-cities/article6380448.ece; accessed on 9 September 2017.

[11] Ravindran, S. 2015. 'Is India's 100 Smart Cities Project a Recipe for Social Apartheid', *The Guardian*, 7 May, available at: http://www.theguardian.com/

One of the major issues is whether smart cities will contribute to the creation of social apartheids in India. Entry into some of these heavily policed cities will be based on the possession of smart ID cards and it is clear that such cities will accentuate the divide between the rich and the poor. Another critical issue is the extent to which privatization will lead to local governance institutions and mechanisms becoming overridden by forms of private governance. One of the most telling visions of smart futures in India was articulated by economist Laveesh Bhandari of Indicus Analytics Pvt. Ltd at a smart cities summit held in Mumbai in January 2015:

> When we build these smart cities we will be faced with a massive surge of people who will desire to enter these cities. We will be forced to keep them out. This is the natural way of things, for if we do not keep them out, they will override our ability to maintain such infrastructure. There are only two ways to keep people out of any space—prices and policing … . Even with high prices the conventional laws in India will not enable us to exclude millions of poor Indians from enjoying the privileges of such great infrastructure. Hence the police will need to physically exclude people from such cities, and they will need a different set of laws from those operating in the rest of India for them to be able to do so. Creating special enclaves is the only method of doing so. And therefore GIFT is an SEZ, and so will each of these 100 smart cities have to be.[12]

The poor in India have now been declared the enemy of PPPs.

Harvey, echoing Marx, explains the ways in which people are dispossessed by processes related to ABD:

> The commodification and privatization of land and the forceful expulsion of peasant populations; conversion of various forms of property rights—

---

cities/2015/may/07/india-100-smart-cities-project-social-apartheid; accessed on 8 September 2017.

[12] Bhandari, L. 2015. 'Smart Cities: What to Do and What Not to Do', in *Smart Cities in India: Reality in the Making,* World Trade Centre, Mumbai, All India Association of Industries, Indo-French Chamber of Commerce and Industry, 29 January, Mumbai. Session 2: Smart Cites & Sustainable Development, p. 16, available at https://www.dropbox.com/s/fr1h3m7d42rnv9a/smart%20cities.pdf?dl=0; accessed on 6 September 2017.

common, collective, state, etc.—into exclusive private property rights; suppression of rights to the commons; commodification of labour power and the suppression of alternative, indigenous, forms of production and consumption; colonial, neo-colonial and imperial processes of appropriation of assets, including natural resources; monetization of exchange and taxation, particularly of land; slave trade; and usury, the national debt and ultimately the credit system.[13]

While the GoI continues to invest in public development expenditures, it is, in this post-liberalization period, also open to the need for private sector investments in financing, operation, and management of e-government projects. There are a number of different contract models of private–public sector partnerships, but among the more common in India are the build–own–operate (BOO) and build–operate–transfer (BOT) models. In some models such as BOT, the projects are designed, constructed, and managed for a finite period, after which assets are returned to the government, whereas in the BOO model, the private firm operates the project on an indefinite basis. There is a major emphasis on the 'efficiency' coefficient in such partnerships—a belief that private sector expertise and skills will lead to a change in the attitudes of public administrators, thereby leading to a change in their delivery of services. The uptake of such models needs to be seen in the context of public sector disinvestments in India and the emphasis on reduced costs, efficiency, and transfer of technologies. While the private sector makes money through payment for transactions, the Indian government has given many of these firms a variety of tax incentives, ranging from import duty exemptions to subsidized land and other incentives. Typically, for example, ICT kiosks and associated front-end activities are financed by firms, while back-end operations, such as telecommunications connections, and so on are supported by the government.

Smart cities conventions and safe city conventions are now commonplace in India. At the International Fire and Security Exhibition and Conference (IFSEC) in 2014, there were numerous product launches, including the following:[14]

[13] Harvey, D. 2004. 'The New Imperialism: Accumulation by Dispossession', *Socialist Register* 40: 63–87.

[14] 'IFSEC 2014', available at http://www.ubmindia.in/IFSEC-2014; accessed on 4 September 2017.

1. *Honeywell Security* launched Pro-Watch 4.2, Armor 300 Wireless Security Kit, HUS-NVR-6032 & HUS-NVR-1032—Honeywell Universal Surveillance, Vista Preconfigured Intrusion Panels suitable for bank branches and automated teller machine (ATM) security application.
2. *Rolta India Limited* showcased its wide range of solutions relevant for homeland security at the exhibition, along with impressive range of Optronics equipment, including day and night vision devices used by security forces.
3. *Matrix* launched complete portfolio of security products encompassing range of access control, time-attendance, and video surveillance solutions.
4. *Hikvision* launched ultra-low-light surveillance DarkFighter PTZ camera.
5. *CP PLUS TeknoLogix Labs* launched composite digital video recorder (DVR).
6. *Silvan Labs* launched the do-it-yourself (DIY) home security systems which enable the user to have a control of their house through an application on their smartphones.
7. *Axis Communications* launched latest intelligent video solutions.
8. *Schneider-Electric* launched Spectra Professional HD PTZ dome camera and Esprit HD PTZ camera.
9. *Sellox B.V. India* launched unique key-centric access management solutions.

India's leading trade security magazine, *a&s India*, also brought out a special issue on city surveillance in India in 2015. An article written by Anant Joshi reported that 28 of the smart cities in India are equipped with a range of surveillance equipment.[15] These include Surat, Mumbai, Pune, and New Delhi. The CCTV surveillance projects based on PPPs have been undertaken in Surat and Mumbai, with the project in Mumbai being undertaken by subsidiaries of Larsen & Toubro, namely, L&T Constructions and L&T Infotech.

All the major software and social networking companies, including Google, Facebook, and Microsoft, have been involved in providing blueprints for 'last-line' solutions linked to rural connectivity projects.

---

[15] Joshi, A. 2015. 'City Surveillance: Booms in India', *a&s India* 50(April): 38.

Google, along with Facebook and Microsoft, are vying to deploy their alternative technologies to offer 'last-mile' broadband connections in remote and inaccessible parts of India to provide access to high-speed Internet. Under an ambitious Rs 1.13 lakh-crore (USD$16.9 bn) 'Digital India' initiative, the government plans to use the national optic fiber network project to deliver e-services in areas such as health, education to every nook and corner of the country.[16]

That Vodafone—the UK-based mobile company that was involved in a non-payment of taxes scandal in the region of Rs 20,000 crores (US$4.3 billion) as recently as 2012[17]—also submitted proposals for the Digital India project does suggest that the State–corporate nexus, even if it is tainted, is fundamental to the workings of the liberal economy.

Surveillance in India, as described in the preceding chapter, is a technology-based management initiative that is aimed at controlling the behaviours and actions of the Indian population. Control, in this instance, is not merely a case of protecting property and people from terrorists and criminals: it also extends to protecting its citizens from pornography and 'talk' that is critical of government ministers and policies. India's Internet filtering agency, the CERT,[18] is involved in both protecting India's cybersecurity and ensuring that security lapses are dealt with. The security and surveillance state is by no means a stand-alone project involving the State. It also involves the private sector. Over the last decade, requests to search engines and ISPs to reveal codes have become commonplace, and the government has made major investments in the interception and interpretation of data traffic. The NSA in the USA hacked into different networks of various countries and was given the information that it

---

[16] Mankotia, A.S. 2014. 'Google in Race with Microsoft, Facebook for a Slice of Digital India', *The Economic Times*, 10 November 2014, available at http://articles.economictimes.indiatimes.com/2014-11-10/news/55955894_1_project-loon-digital-india-facebook-and-microsoft; accessed on 9 September 2017.

[17] Siva, M. 2014. 'All You Wanted to Know about Vodafone's Tax Case', *The Hindu Business Line*, 17 February, available at http://www.thehindubusinessline.com/opinion/all-you-wanted-to-know-about-the-vodafone-tax-case/article5699526.ece; accessed on 12 September 2017.

[18] DeitY, 'Computer Emergency Response Team', available at http://deity.gov.in/content/icert; accessed on 11 January 2018.

required by major search engines such as Google, social networking sites such as Facebook, and firms such as Apple. One of the more contentious cases regarding such requests was related to the Canadian mobile phone company Blackberry and its initial refusal to divulge its code related to its instant messaging services, followed by a retreat in the light of the government's threat to shut down its services.[19]

Control via surveillance is, however, only one aspect of the equation; the other is control via the government's extension of welfare. In this instance, code ostensibly plays a more benign role ensuring that the government's online functionalities are executed, services delivered efficiently, and delivery interfaces facilitate access and data flows. However, and arguably, e-governance can also be seen as an attempt to manage populations efficiently, especially in the provisions of welfare through multiple techno-managerial strategies. The Aadhaar scheme, discussed in Chapter 1, is an example of welfare-mediated services whose aim is to provide each resident Indian with an identity that can be called upon in the disbursement of welfare entitlements, and that also functions as an ID. Apart from the government that has substantive interests in code as control, there are also key actors in the private sector for whom code is the basis for controlling and deepening the consumption habits of consumers through fine-tuning consumer choice. This chapter will explore both types of control in which the substantive rights of citizens have either been compromised by Big Brother or by key actors in the consumer economy such as Google and social networking sites such as Facebook that are involved in selling consumers to advertisers. In both cases, code plays an important role in the structuring of options.

The Surveillance of 'Affect'

From the perspective of leisure, the power of algorithmic code to structure affective behaviour, relevance, rank, and visibility remains the key issue—an issue that has been amplified in the context of platform availability on a number of consumer devices, from mobile phones to iPads and tablets. The availability of search engines and social networking sites across multiple personal devices has led to industry investments in

[19] Ziccardi, G. 2012. *Resistance, Liberation Technology and Human Rights in the Digital Age*, The Netherlands: Springer Science and Business Media, p. 287.

predictive software, along with the development of an industry that is involved in the tracking and mining of personal information. This tracking of personal information remains an extraordinarily opaque area that is beyond the scrutiny of individual users involved in generating their own data, precisely because the act of Googling or Facebooking has become a routine aspect of the ways in which people communicate in the twenty-first century. The normalization of social network sites—that are each involved in the curation, mining, and business of personal information—is an extraordinary aspect of contemporary life.

Code as Power

From a political economy perspective, we take code and algorithmic code seriously precisely because it has become the means to enforce extraordinary power on a global scale. This power is a direct consequence of the digitalizing of everything that has become the basis for global capitalism today and is one in which informational goods and services, as well as processes and products have become fundamental to the enactments of public and private communications. When productive functions become dependent on the activations of code, there are possibilities for power to be concentrated in enterprises such as social networks that are intentionally global, networked, and dependent on user connectivities, and search engines that are likewise dependent on users as consumers of information in all its variety. This power, as Scott Lash has pointed out, is different precisely because it acts on an immanent reality—the fact that the digital is ubiquitous and, like the ether, envelops us within its normalcy: 'Pouvoir was much easier to unmask when it worked from the outside as power-over. The critique of ideology of left-hegemonic politics could manage this. But when power enters into us and constitutes us from the inside … it becomes far more difficult to unmask.'[20] This power, according to Lash, is based on a new set of rules based on code. These are not 'constitutive' or 'regulative' rules that have traditionally structured societies,[21] but are "'generative" rules … virtuals

---

[20] Lash, S. 2007. 'Power after Hegemony: Cultural Studies in Mutation', *Theory, Culture & Society* 24(3): 61.

[21] Lash, 'Power after Hegemony', p. 70.

that generate a whole variety of actuals. They are compressed and hidden and we do not encounter them in the way that we encounter constitutive and regulative rules. ... They are ... pathways through which capitalist power works ... a society of ubiquitous media means a society in which power is increasingly in the algorithm.'[22] However, unlike Lash who seems to believe that in an era in which power is self-produced, all previous exercises of power and modalities have been superseded, I argue that old and new types of power coexist and feed off each other. We are not in a post-hegemonic era precisely because direct flows of power continue to dictate the nature of war and peace as much as the nature of trade. After all, the power of a Rupert Murdoch or Google is linked to the fact that New Corporation was worth $39.6 billion in 2011 and Google's market capitalization in 2014 was $395 billion, next only to Apple that was worth $465 billion and the most valuable company in the USA.[23] In other words, it is important that as we look at how value is generated by social networks and search engines, we focus on the role played by individuals as they enter into productive relationships with these networks—and in particular, how their multiple connectivities contribute to the actual value that is generated by networks who in turn sell information generated through such transactions to advertisers and data-mining companies. This points to the concept of the 'audience-commodity' that the erstwhile political economist Dallas Smythe first explored[24]—the fact that the key function of broadcasters is to sell audiences to advertisers. That reality has not changed and, in fact, has been perfected in the age of social networks and search engines that have perfected the art of selling individuals' wants and needs to the market.

[22] Lash, 'Power after Hegemony', p. 71.

[23] The power of a Bill Gates, Jeff Bezos, and other owners of companies in the digital economy is reflected in their power to set agendas and in the market capitalisation of their companies in 2018: Microsoft ($753 billion), Amazon ($782 billion), Alphabet ($739 billion), Apple ($923 billion) (See Warren, T. 2018. 'Microsoft Is More Valuable than Google', *The Verge*, 30 May, available at: https://www.theverge.com/2018/5/30/17408254/microsoft-google-alphabet-market-cap-value; accessed on 13 March 2019.

[24] Smythe, D. 2006. 'On the Audience Commodity and Its Work', in M.G. Durham and D. Kellner (eds), *Media and Cultural Studies: Key Works*. Malden, Oxford, and Carlton: Blackwell, pp. 230–56.

Algorithmic power is intriguing because it is hidden behind a discourse of information for all, of access, of the semantic Web, and of the rhetoric linked to the democratization of the Web. And yet, it is fundamentally built on an old paradigm, that of 'property'. Property values are exuded in every bit of code, which is why algorithms within the capitalist framework cannot be made transparent or given away for free simply because it is the basis for competitive advantage and profit margins. The WikiLeaks files on the status of IP in the Trans-Pacific Partnership Agreement, the world's largest trade agreement, include an entire chapter on stronger IP protection and enforcement.[25] Dorling, writing in *The Sydney Morning Herald* on its implications for the Australian consumer, highlights the fact that: 'Intellectual property experts are critical of the draft treaty, which they say would help the multinational movie and music industries, software companies and pharmaceutical manufacturers to maintain and increase prices by reinforcing the rights of copyright and patent owners, clamping down on online piracy, and raising obstacles to the introduction of generic drugs and medicines.'[26]

Code is only given away for free within a paradigm that is explicitly linked to the creation of alternatives, such as FOSS. So, in this sense, algorithms are mainly built on the foundations of old capital. What is new is the way in which it is used to create value through the labour of individuals via their use of search engines and social networks. Its newness is profoundly different from earlier ways of creating value precisely because it is based on harvesting and encouraging tastes based on user profiles. This mining is a function of algorithmic power and the reason for its extraordinary successes. Hallinan and Striphas explain the nature of what they have termed "algorithmic culture": provisionally, the use of computational processes to sort, classify, and hierarchize people, places,

[25] 'Updated Secret Trans-Pacific Partnership Agreement (TPP)—IP Chapter (second publication)', available at https://wikileaks.org/tpp-ip2/#article_gzz; accessed on 11 February 2018.

[26] Dorling, P. 2013. 'Australians May Pay the Price in Trans-Pacific Partnership Free Trade Agreement', *The Sydney Morning Herald*, 14 November, available at http://www.smh.com.au/federal-politics/political-news/australians-may-pay-the-price-in-transpacific-partnership-free-trade-agreement-20131113-2xh0m.html; accessed on 13 February 2018.

objects, and ideas, and also the habits of thought, conduct, and expression that are in relationship to those processes.'[27]

Understanding Code

The materiality of the information structures that surround us is intricately tied to and purposed by the conditionalities of algorithmic code. Code animates all information structures, from the mundane such as the software that runs cars, refrigerators, and mobile phones to search engines and social networks, manufacturing and telecommunications, military operations, and the financial sector. The malfunctioning of code can have both mildly irritating and catastrophic consequences—a faulty toaster/burnt toast as well as a faulty missile/collateral damage. Code is always in the background and is seldom visible precisely because it is very much like the power that runs batteries and lights up our houses. However, unlike electricity, code is a language, it is a property, and hence is material, although by its very nature the operationalization of code is intangible. Martin Dodge explains the relationship:

> Conceptually software is built of lines of code—simple instructions and algorithmic rules—that when combined together with appropriate data produce operative programs capable of complex functions. Software could be thought of [as] a special kind of written language, with particular grammatical rules, vocabularies and linguistic conventions. Rather than being printed and read by people, code *runs*, it is a *self-executable* language (emphasis in original).[28]

Algorithms animate functions in the digital worlds that we inhabit. One can, however, argue that the algorithm has become key to digital capitalism, given that it acts as the fuel that enables a range of operations in the economy, politics, culture, and society. Totaro and Ninno have suggested that algorithms, fundamentally, are about the execution of functions (algorithms manipulate non-numerical objects) that are key to the logic and workings of informational capitalism. In their words:

---

[27] Hallinan, B. and T. Striphas. 2016. 'Recommended for You: The Netflix Prize and the Production of Algorithmic Culture', *New Media & Society* 18(1): 119.

[28] Dodge, M. 2009. 'Code/Space', *Urbis Research Forum Review* 1(2) 2009: 16.

'the logic of algorithms (i.e. recursive function) is the specific form in which the concept of function occurs in everyday life as is the case with the calculation of real economic transactions or with algorithms used for the production and provision of many goods and services.'[29] While it is important to explore the role played by code, to overstate its specificity and importance at the expense of the larger political economy of digital capitalism remains an ever-present problem for scholars of code. In the words of Hillis, Michael, and Jarrett:

> Attributing causal agency to algorithms that are designed by engineers, moreover, works to sever these necessarily ideological decisions from the broader institutional and socio-economic settings in which such decisions are naturalized and which in the first place have led to the production of algorithms that function in particular ways. Code has important ideological effects but it is equally important to recognize that code itself is an ideological effect.[30]

And yet, it is arguable that both the technocratic imaginary and the functionalities that algorithms help execute and deliver, inclusive of ubiquitous surveillance and expansive techno-management of welfare through e-governance, render it a subject that is worth exploring from a political-economy perspective.

The Predictive Power of Algorithms

Algorithms have not only become the basis for precise predictions in business but also a means of taking evasive action, that is, controlling potentially trouble-making populations through profiling. Algorithms, in the context of enforcing security, can result in recursive (repetitive) solutions and the routine incarcerations of people precisely because of their congregating at a neighbourhood that has had a history of anti-social behaviour. The Criminal Reduction Utilizing Statistical History (CRUSH) is based on IBM's predictive analysis software and is used

---

[29] Totaro, P. and D. Ninno. 2014. 'The Concept of Algorithm as an Interpretative Key of Modern Rationality', *Theory, Culture & Society* 31(4): 32.

[30] Hillis, K., M. Petit, and K. Jarrett. 2012. *Google and the Culture of Search*. Hoboken: Taylor & Francis, p. 21.

in the USA for policing operations, such as by the police department in Memphis, Tennessee.[31] Here, the intent is to create profiles of neighbourhoods and incident hot spots to create proactive evasive action through clamping down on individuals whose profile suggests a potential risk to the neighbourhood. As the director of police in Memphis, Toney Armstrong, explains:

> Shortly after Blue CRUSH was announced in 2005, MPD conducted a series of test pilot operations in selected precincts before rolling out the program city-wide. We used data analysis of past and current crime information provided with IBM SPSS predictive analytics software to evaluate incident patterns throughout the city—in areas as wide as the city's entire nine precincts or narrowed down to a single block. Maps and charts of crime patterns were generated based on type of criminal offense, time of day, day of week or various victim/offender characteristics. These maps were then used to specifically focus investigative and patrol resources with the goal of taking back neighborhoods one street at a time. The initial results were staggering and quickly validated our strategy.[32]

In this particular case, code has become the basis for enforcing security, and reinforces and extends the dominant ideology of policing and solutions.

What the two given examples demonstrate is that algorithms have become the basis for the normalization of productive activities and the shaping of spaces and societal expectations, such as controlling 'risk' by pre-empting violence on the streets. In both cases, the mitigation of risk offers opportunities for greater productivity, well-being, and security—fundamental to the contemporary ideology of capitalism. Kitchin and Dodge refer to the discourses that the state bureaucracies and corporations use to invite the complicity of audiences who are willing to trade 'potentially disciplinary effects against benefits gained', including 'safety, security, efficiency, anti-fraud, empowerment, productivity,

---

[31] Utsler, J. 2011. 'The Crime Fighters', *IBM Systems Magazine*, February, available at http://www.ibmsystemsmag.com/power/trends/ibmresearch/ibm_research_spss/.

[32] Armstrong, T. 2013. 'Managing for 21st Century Crime Prevention in Memphis', *Management Innovation Exchange*, 7 January, available at http://www.managementexchange.com/story/managing-21st-century-crime-prevention-memphis; accessed on 12 September 2017.

reliability, flexibility, economic rationality, and competitive advantage.'[33] In other words, algorithms now facilitate the making of an ideal society in which any 'difference' that can impact the production–consumption compact has been edited out. That would include those who do not have the capacity to consume, those who are differentially placed on the socio-economic ladder, and, in the context of the USA and the example from Memphis, entire neighbourhoods that are racially segregated and in which racial profiling is shaped by algorithms.

Algorithmic functionalities, in other words, contribute to the widening of pre-existing divides, although that in itself, strictly speaking, does not make it ideological. It does, however, highlight the nature of code and the choices that human beings make to make code function to specific ends. Those ends, arguably, are ideological in nature. Lawrence Lessig captures the dilemma of code and locates it in human agency: 'code presents the greatest threat to liberal or libertarian ideals, as well as their greatest promise. We can build, or architect, or code cyberspace to protect values that we believe are fundamental, or we can build, or architect, or code cyberspace to allow those values to disappear. There is no middle ground.'[34] In other words, algorithms and codes are socially constructed. Mager has explored the notion of an 'algorithmic ideology'.[35] However, strictly speaking, the algorithm is not an ideology like capitalism is. What it does is encapsulate the key means by which functions are executed digitally/ recursively across a whole range of productive activities, across multiple sectors in the bureaucracy, the life sciences, the leisure industries, and other sectors that are involved in the business of growth and production that are key to capitalism. Algorithms and code make sense in the context of software programmes that are designed to execute specific purposes. As Cheney-Lippold explains:

[33] Kitchin, R. and M. Dodge. 2011. *Software Studies: Code/Space: Software and Everyday Life*. Cambridge, MA: MIT Press, p. 106.

[34] Lessig, L. 1999. *Code and Other Laws of Cyberspace*. New York: Basic Books, p. 6.

[35] Mager, A. 2012. 'Algorithmic Ideology', *Information, Communication & Society* 15(5): 769–87; and Mager, A. 2014, Defining Algorithmic Ideology: Using Ideology Critique to Scrutinize Corporate Search Engines', *tripleC* 12(1): 28–39.

The study of code and software in general tries to go 'beyond the blip' to understand the implicit politics of computer code, to make visible the dynamics, structures, regimes and drives of the wide variety of programmed scripts that are littered across the internet. With this view we can see computer code not as a literal, rhetorical text and from which we can derive meaning but a complex set of relationships that tie together the coded systems of definition and organisation that constitute our experiences online. Codes are cultural objects embedded and integrated within a social system whose logic rules, and explicit functioning works to determine the new conditions of possibilities of user's lives.[36]

Code and the Manipulation of Sentiments

This State–corporate nexus is hugely problematic given that the two impact people's lives in so many different ways, irrespective of where one is geographically located. While there are many permutations of this nexus, their joint impact on the lives of citizens can be quite extraordinary. There is also a sense in which the terms used to describe what Google does, that is, 'search', and what Facebook does, that is, 'social networking', are misnomers precisely because both terms have been manufactured. The dictionary meanings of these terms certainly do not suggest that our 'searches' and 'social networking' contribute to our own surveillance and to the creation of profiles that are, in turn, available for sale to the State and the market. So, the very normality of these terms, common sense in the Gramscian sense, has contributed to the making of information hegemons such as Google. Algorithms therefore, first and foremost, contain an implicit politics because, as a programme, they can and do structure options and the boundaries for searchability and social networkability. Second is the seeming willingness by users to cede control over their private lives and information to social networks. There are different aspects to this type of hegemony. Users are willing to commit themselves to online sociality via multiple connectivities and are, for the most part, oblivious to the reality of data mining. In fact, such realities do not figure in the average user's uses of social networking, and even if they do, this is not considered a serious breach of privacy as most users are

---

[36] Cheney-Lippold, J. 2011. 'A New Algorithmic Identity: Soft Biopolitics and the Modulation of Control', *Theory, Culture & Society* 28(6): 167.

yet to experience threats related to data mining that have impacted their lives. Users seem willing to forgo freedoms for 'reason of security' as long as they can get on with their connectivities. This remains a grey area as demonstrated by the following example.

An issue of the *Proceedings of the National Academy of Sciences of the United States of America* (PNAS) published a study of an experiment done by the Core Data Science team of Facebook with the Departments of Communication and Information, Cornell University, in which they used Facebook user (n = 689,003) data manipulation to find out whether 'emotional states can be transferred to others via emotional contagion' without the users being aware of it.[37]

> In an experiment with people who use Facebook, ... [the authors tested] whether emotional contagion occurs outside of in-person interaction between individuals by reducing the amount of emotional content in the News Feed. When positive expressions were reduced, people produced fewer positive posts and more negative posts; when negative expressions were reduced, the opposite pattern occurred. These results indicate that emotions expressed by others on Facebook influence our own emotions, constituting experimental evidence for massive-scale contagion via social networks. This work also suggests that, in contrast to prevailing assumptions, in-person interaction and nonverbal cues are not strictly necessary for emotional contagion, and that the observation of others positive experiences constitutes a positive experience for people.[38]

While the study reveals the potential for Facebook to be manipulated for propaganda purposes, the team did not take prior informed consent from the users because all users who have accounts on Facebook automatically comply with their Facebook Data Use Policy.[39] Their use of user pages does go against their own policy that states:

[37] Kramer, A.D.I., J.E. Guillory, and J.T. Hancock. 2014. 'Experimental Evidence of Massive-Scale Online Contagion through Social Networks', *PNAS* 111(24): 8788.

[38] Kramer, Guillory, and Hancock. 'Experimental Evidence of Massive-scale Online Contagion through Social Networks', p. 8788.

[39] Facebook. 'Data Use Policy', available at https://www.facebook.com/full_data_use_policy; accessed on 12 September 2017.

While you are allowing us to use the information we receive about you, you always own all of your information. Your trust is important to us, which is why we don't share information we receive about you with others unless we have:

- received your permission;
- given you notice, such as by telling you about it in this policy; or
- removed your name and any other personally identifying information from it.[40]

The third way in which the impact is felt concerns specific algorithms, such as EdgeRank (Facebook) and PageRank (Google), that are used to establish visibility, ranking, and greater correspondences between users and the market.

Leisure and Code

It is in the business of leisure that algorithms have become the basis for affecting the nature of choices, options, and the very structuring of possibilities. It is this aspect that has received the largest number of contributions in journals like *Theory, Culture & Society*, *Information, Communication & Society*, *New Media & Society*, and *Surveillance & Society*. One of the central themes in these writings is that of information surveillance and the need to understand algorithmic power in the context of the production–consumption–surveillance axis. The politics of 'big data' and data mining is another theme that has attracted attention in the context of the growth of search engine giants, social networks, and their complicity with the NSA in the USA in the aftermath of the post-Snowden revelations. In the context of algorithms in leisure, the intent is to precision calibrate the production–consumption compact, thus increasing possibilities for complete correspondences between individual leisure choices, such as the choice of a film for viewing and the film itself which has been tailor-made for box-office success.

It is in the context of leisure and the use of the technologies of everyday life—search engines and social networks—that the algorithm takes on special significance. There are a number of vantage points that one can use to apprehend and make sense of algorithms. First, there is a

---

[40] Facebook. 2014. 'Data Use Policy'.

need to understand the algorithms that animate predictive filtering and ranking devices used by Google, Facebook, and other leisure outlets; second, to understand how such mechanisms work in practice; and third, to understand its consequences for both consumers and the owners of platforms.

Epagogix, Hollywood, and Bollywood

Hollywood studios routinely use Epagogix, an algorithm-driven software that is used to predict the box-office viability and success of film scripts. Given that investments in films run into millions of dollars, box-office failure can have a crippling impact on studios. As it is stated on their website: 'Advanced Artificial Intelligence in combination with proprietary expert process enables Epagogix to provide studios, independent producers and investors with early analysis and forecasts of the Box Office potential of a script. Clients then make evidenced decisions about whether or not to spend their scarce capital, adjust budgets, or to increase the Box Office value of the property.'[41] Their predictive software enables film and TV studios to reduce risk: 'Epagogix works confidentially with the senior management of major film studios, large independents and other media companies, assisting with the *selection and development of scripts* by identifying likely successes and probable "Turkeys"; helping to quantify a script/project's commercial success; and advising on enhancements to the Box office/audience share potential.'[42] Epagogix software is based on a crunching of big data on box-office takings and audience surveys, to predict the success of a film script, along with the optimum role characterizations for and between roles and actors. As Philip Napoli has observed: 'the more motion picture studios depend on Epagogix to determine their production slate, the more likely it would seem that these studios will produce similar films, as the same algorithms and underlying data are driving their decision making. In this regard, algorithmically driven institutional isomorphism essentially results in

[41] 'Epagogix', available at http://www.epagogix.com/; accessed on 12 September 2017.

[42] Fitzgerald, E. 2017. 'The Algorithm Will Read Your Story Today', *Homer*, 15 February, available at http://www.eamonn.com/2017/02/15/the-algorithm-will-read-your-story-today/; accessed on 13 March 2019.

diminished diversity of content output.'[43] If algorithms are used to predict the success of a script, then the formulaic film will inevitably triumph over films that have novel plot lines and that are based on idiosyncratic stories. However, it is more than just the types of films in circulation that will be affected—for it is fundamentally and potentially software that can affect and shape what is available to us as public culture. Is this something that we need to be worried about? While there will always be the space for films that are different to the norm, big studios that are dependent on box-office revenues might opt for the 'popular' as opposed to the 'creative' and/or 'challenging'. Arguably, predictive software is merely narrowing options even further in a market that began that process decades ago. In 2007, Netflix, the on-demand Internet-streaming company that led to the demise of the Blockbuster video outlet in the USA and elsewhere, had invested in a $1 million prize for someone who could develop a system that could predict movie ratings. Katie Hafner, writing in *The New York Times*, states: 'About 18,000 teams from more than 150 countries—using ideas from machine learning, neural networks, collaborative filtering and data mining … submitted more than 12,000 sets of guesses. And the improvement level to Netflix's rating system is now at 7.42 percent.'[44]

The fact that Bollywood too is now using predictive softwares to ensure box-office success is to be expected given the scale of the movie industry in India. The failure of big-budget films, such as South Indian superstar Rajnikanth's film *Lingaa* in 2015, arguably hastened the use of such softwares. In 2014, IBM introduced its predictive software into the Indian film market that was developed after 'IBM analysed over seven lakh posts across a variety of platforms including Facebook, Twitter, YouTube and blogs for 25 Bollywood films with considerable social buzz. These included films like Ek Tha Tiger, Barfi, Bhaag Milkha Bhaag, Kai Po Che, Kahaani and Agneepath.'[45] Using IBM's Social Sentiment Index,

[43] Napoli, P.M. 2014.'Automated Media: An Institutional Theory Perspective on Algorithmic Media Production and Consumption', *Communication Theory* 24(3): 352.

[44] Hafner, K. 2007. 'Netflix Prize Still Awaits a Movie Seer', *The New York Times*, 4 June, available at http://www.nytimes.com/2007/06/04/technology/04netflix.html?_r=0; accessed on 8 September 2017.

[45] John, S. 2014.'IBM Brings in Analytics to Bollywood', *The Times of India*, 17 March, available at http://timesofindia.indiatimes.com/tech/it-services/IBM-brings-in-analytics-to-Bollywood/articleshow/32196946.cms; accessed on 9 September 2017.

mathematical and probability modelling, and machine learning, they have attempted to define the hit film and have suggested that political themes translate into box-office success. Predictive analytics is also being used in the TV industry in India to increase the brand value and brand identification, and predict the salience of content dealt with in an investigate talk show, *Satyamev Jayate* (Truth will Triumph), which has dealt with a range of social issues in India, including female foeticide, child sexual abuse, and medical malpractice, among other issues.

EdgeRank, PageRank, and the Role of Algorithmic Code in the Leisure Industry

EdgeRank is an algorithm that was created by Facebook to determine and display ranked material in a user's newsfeed. The algorithm is based on three edges: affinity, weight, and time decay. These edges refer to all activities within Facebook; in other words, the entire gamut of user interactivity on Facebook, based on the means of interactivity available. Affinity is calculated by a user's closeness to a brand or product based on the frequency of user interactivity and use of Facebook's modes of interactivity, including sharing, liking, and so on.

> Facebook calculates affinity score by looking at explicit actions that users take, and factoring in 1) the strength of the action, 2) how close the person who took the action was to you, and 3) how long ago they took the action. Explicit actions include clicking, liking, commenting, tagging, sharing, and friending. Each of these interactions has a different weight that reflects the effort required for the action—more effort from the user demonstrates more interest in the content. Commenting on something is worth more than merely liking it, which is worth more than merely clicking on it. Passively viewing a status update in your newsfeed does not count toward affinity score unless you interact with it. Affinity score measures not only my actions, but also my friends' actions, and their friends' actions. For example, if I commented on a fan page, its worth more than if my friend commented, which is worth more than if a friend of a friend commented. Not all friends' actions are treated equally. If I click on someone's status updates and write on their wall regularly, that person's actions influence my affinity score significantly more than another friend who I tend to ignore.[46]

---

[46] 'EdgeRank', available at http://edgerank.net/; accessed on 6 September 2017.

Weight is calculated on relative weighting given to these modes of interactivity. For example, commenting is given more weight than liking. And time decay is based on the relative salience of the edges. The older the edge, the less relevant it becomes. Given the extraordinary opportunities today for brands to extend their market value through Facebook pages, it is clear that brands invest in maintaining a steady flow of content.

However, EdgeRank affects all Facebook users, including individuals. EdgeRank ranks the content in user feeds and ensures that only those who relate to your views are part of your network. At the same time, those who have antagonistic or opposed views on the nature of politics in any given country on their Facebook page can become a victim of trolls backed by dominant politics. Given the fact that all Facebook conversations are archived, conflictual views that are contrary to the mainstream can be publicly exposed and this can invite both political and judicial scrutiny. Andrejevic has observed that while Facebook offers all those who have Facebook pages to use it for activist purposes, 'the move from marketing to social activism is neither simple nor automatic. … it comes about through engaged social struggle rather than sanguine reliance on the character of the technology.'[47] In other words, sites such as Facebook use code in ways that privilege the activities of the market and the type of interactivity necessary for brands to grow, rather than sites that are based on 'agonism' and whose content is different from the mainstream. In this sense, EdgeRank restricts the visibility of certain types of content, while it encourages the visibility of other types of content. Thus, in this regard, we need to take Bucher's observations on EdgeRank seriously. He observed: 'The algorithm is not only modelled on a set of pre-existing cultural assumptions, but also on anticipated or future-oriented assumptions about valuable and profitable interactions that are ultimately geared towards commercial and monetary purposes.'[48] Individual predispositions are reinforced in other ways through the architecture of code and its organizational methods, resulting in group conformity and an endless communication among the converted. The social consequences of such segmented behaviour are manifold, given that Facebook and other social networking sites offer opportunities for us

---

[47] Andrejevich, M. 2007. *iSpy: Surveillance and Power in the Interactive Era.* Lawrence, Kansas: University Press of Kansas, p. 50.

[48] Bucher, T. 2012. 'Want to Be Top? Algorithmic Power and the Threat of Invisibility on Facebook', *New Media & Society* 14(7): 1169.

to find solace and comfort among the like-minded. Rather paradoxically, they could be like-minded racists or like-minded progressives. One can argue that such sites do not, therefore, contribute or augment any serious dialogue between civilizations given that the code structures communication among the converted. Evidence related to the potential for the spread of emotional contagion via social networks, referred to earlier in this chapter, merely confirms the view that public policies need to be used to regulate the behaviour of social networks involved in the business of private information.

Google's PageRank algorithm is similar to Facebook's EdgeRank in that it ranks traffic flows and search results. Arguably, given Google's extraordinary power as a global information giant, its search algorithm is part of an extensive stable of search engines that it owns and that is involved in mining and matching user interests to the market. Moreover, PageRank is the most important among the algorithms owned by Google. It is an evolving, ever-changing algorithm given that 500–600 modifications and changes are made to it every year.[49] Christian Fuchs' article on the political economy of Google is one of the most comprehensive and critical pieces on the surveillance power of this global information giant.[50] Fuchs provides a typology of Google surveillance based on the many ways in which Google applications carry out a surveillance of one's personal identity and economic data, the nature of data collected, and the devices/platforms used—from Gmail to Google's mobile applications, YouTube videos, etc.[51] Fuchs highlights 10 issues related to Google that have attracted attention from academics, including monopolization of the search engine market, surveillance, censorship, and its political dominance, among other issues. Mager specifically comments on Google's PageRank: 'Systematically preferring big, well-connected websites at the expense of smaller ones', thus curbing 'the democratic potential of the web'.[52] But there is more to PageRank than its ranking system or its potential for

[49] Google Algorithm Change History, Moz. n.d., available at https://moz.com/google-algorithm-change; accessed on 13 March 2019.

[50] Fuchs, C. 2011. 'A Contribution to the Critique of the Political Economy of Google', *Fast Capitalism* 8(1): 1–28.

[51] Fuchs, 'A Contribution to the Critique of the Political Economy of Google', pp. 16–18.

[52] Mager, 'Defining Algorithmic Ideology', p. 29.

dataveillance. Critically, PageRank helps categorize and mine big data sets and search engine–based expressions of social desire. The mining of these data sets provides Google with information on the attentive/cognitive behaviour of individuals and collectivities online. This information, in turn, becomes the targeted focus for its advertisement platform, AdSense, that:

> provides a light infrastructure for advertising that infiltrates each interstice of the web as a subtle and mono-dimensional parasite, extracting profit without producing content. … Within the economy of the Internet, both the traffic of a website and the redistribution of value are extensively governed by *PageRank*. *PageRank* is at the core of the attention economy of the Internet.[53]

The Double Movement

While Polanyi described the double movement in terms of State and society-based efforts to bring a semblance of social equilibrium to the market, in the context of the twenty-first century, it is arguably the case that civil society plays an important role in humanizing the market. The FOSS movements are, for example, involved in resisting the tyranny of proprietorial code. Lessig has observed that the 'regulation of behavior in cyberspace … is imposed primarily through code. What distinguishes different parts of cyberspace are the differences in the regulations effected by code … Some architectures of cyberspace are more regulable than others, some architectures of cyberspace enables [sic] better control than others … its architecture is its politics.'[54] Lessig's critique was focused on the role of big government in regulating code and less on the role played by the private sector.

Kelty has argued that 'the "reorientation of power and knowledge" has two key aspects that are part of the concept of recursive publics: availability and modifiability (or adaptability)'. While 'availability' refers to the need for environments that are supportive of principles such as

---

[53] Pasquinelli, M. 2009. 'Google's PageRank: Diagram of the Cognitive Capitalism and Rentier of the Common Intellect', in Konrad Becker and Felix Stalder (eds), *Deep Search: The Politics of Search beyond Google*. Innsbruck, Wien, and Bozen: StudienVerlag, p. 156.

[54] Lessig, *Code and Other Laws of Cyberspace*, p. 20.

'transparency, open governance or transparent organisation, secrecy and freedom of information, and open access in science ... "Modifiability" refers to the right to not only access software and hardware but to also adapt it, transform it to suit local needs, along with the right to circulate and redistribute this material. An example of the translation of this right are the provisions for modifiability included in alternative intellectual property licensing arrangements such as Creative Commons, although arguably recursive publics contribute to both minimalist and maximalist versions of these rights.'[55]

There has certainly been a slow but steady growth in FOSS and the ecosystem needed to support FOSS in India. The central government as well as various state governments have invested in FOSS and there are a number of FOSS initiatives in the country supported by the State and civil society that involve consumers, producers, and facilitators of FOSS.

Code and Community in India

One of the better examples of the State–civil society efforts to redeem code as a public good for public use is the IT@School project, a partnership between the southern Indian state of Kerala and the NGO, the Society for Promotion of Alternative Computing and Employment (SPACE). Since 2005, SPACE has been involved in the production of free-software based solutions for e-government, in public education and for marginalized citizens such as the visually disabled. To a large extent the success of public software can be attributed to bi-partisan political support and the active presence of software activists. The FOSS-based IT@School project in Kerala has, for example, led to ICT-enabled education in 8000 schools in the state and has involved the training of 200 master trainers and 5600 IT coordinators (selected from among the teaching staff).[56] With a strong emphasis on e-education and use of FOSS for employment, the objectives of SPACE are as follows:

[55] Kelty, C.M. 2008. *Two Bits: The Cultural Significance of Free Software.* Durham & London: Duke University Press; Kelty, C.M. 2005. 'Geeks, Social Imaginaries and Recursive Publics', *Cultural Anthropology*, 20(2), pp. 185–214.

[56] See Thomas, P.N. 2011. *Negotiating Communication Rights: Case Studies from India.* New Delhi: SAGE, p. 185.

1. Advocate multi-purpose, cross-sectoral uses of FOSS across the public and private sector.
2. Provide institutional support and enabling environments for FOSS-based training and business opportunities.
3. Invest in R&D related to FOSS
4. Invest in the human resources required to strengthen the imprint of FLOSS across the nation.[57]

SPACE has been involved in both making available and modifying platforms and operating systems, in particular, for marginalized communities such as the visually disabled. They have pioneered the use of assistive technologies based on combinations of speech synthesis technology, braille, and magnification to create a powerful aural tool for the visually impaired (via a program called Insight) that is complemented with an audio magazine, *Swaram*. A related organization, the multistakeholder International Centre for Free and Open Source Software (ICFOSS) established by the Government of Kerala in 2011 was responsible for launching the world's largest FOSS-based IT facility, 'Swatantra', in the government sector in January 2019. It will include FOSS training spaces and extend capacity training initiatives.[58]

Given India's strength in programming, Microsoft's decision to work on advanced software algorithms in its India office does not come as much of a surprise. While proprietary algorithms, especially algorithms to help brokers in the derivatives market, attract major investments, the fact that India does have a sizeable number of programmers committed to FOSS-based solutions suggests that civil society will have a stake in the development of another algorithmic culture. This is certainly the case in the IT@Schools project in Kerala.[59] As mentioned earlier, to some extent Kerala has been ahead of other states in India in the adoption of FOSS due to bipartisan political support. In an interview

[57] Thomas. *Negotiating Communication Rights: Case Studies from India*.

[58] Dineshwari, L. 2019, 'Kerala Houses World's Largest Free and Open Source Facility "Swatantra"', *OpenSource*, 2 March, available at https://opensourceforu.com/2019/03/kerala-houses-worlds-largest-free-and-open-source-facility-swatantra/; accessed on 13 March 2019.

[59] See Thomas, P.N. 2014, 'Public Sector Software and the Revolution: Digital literacy in Communist Kerala', *Media, Culture & Society* 36(2): 258–68.

in 2010, Sasi Kumar from SPACE described the history of FOSS in the state as follows:

> One thing in Kerala was that IT education started early. IT as a subject and multimedia presentations, spreadsheets, word processing were taught. No other state in India had such a policy. Initially it was based on Windows and Microsoft Office. However there was opposition from the FOSS crowd in Kerala who were of the opinion that the government should not promote products that belong to any particular company. When the Left Front came to power they adopted an IT policy that was drafted with inputs from the FOSS community. It states that IT is not only for industry but that all software used for pubic should be FOSS-based.[60]

The Kerala government has shown its support for FOSS through other initiatives as well, including hosting the 'Freedom First' conference in 2001 at which Richard Stallman was the chief guest. It has also invested in the Centre for Advanced Training in Free and Open Source Software. In other parts of India too, such as in Karnataka, NGOs like IT for Change have played a key role in introducing FOSS to the public sector. There have also been prominent civil servants who have acted as evangelists for FOSS and central government investments in initiatives such as the National Resource Centre for Free/Open Source Software (NRCFOSS) that has been instrumental in creating BOSS that is the basis for much of the government-sponsored FOSS initiatives. So, in the Indian case, FOSS as a movement has been supported by the State, the civil society, and the industry.

It is clear that in the Indian context, code as a public good is primarily directed towards ensuring that the majority of Indians are digitally enabled to access and make use of online welfare, education, health, and other services. While there are obvious benefits for citizens, there are also risks associated with the use of such personal data by the State—risks that, in the absence of strong privacy laws and the presence of a Hindu-nationalist government, are pronounced. There has been very little civil society advocacy related to contesting the use of algorithmic code by the private sector to manipulate individual choices and preferences of Indians online. The battle though is not just about ensuring access to code but it is also

---

[60] Sasi Kumar, Personal interview, Bengaluru, India, 26 February 2010.

about the need to break current monopolies in search engines and social networking. As this chapter has highlighted, the State–private sector nexus has led to a situation where ordinary Indians as citizens and consumers are susceptible to the influence of code in their private and public lives.

The announcement by the GoI that open source software will be the basis for the Digital India project[61] does suggest that even the current, pro-market government has recognized the value of publicly available code—thus contributing to the double movement described by Polanyi. However, such accommodations need to be seen in the light of the embrace of Google and Facebook by the GoI and the involvement of these organizations in the Digital India project. Mark Zuckerburg's Internet. org project is aimed at bringing the Internet through mobile applications such as the mobile phone to a billion people, including Indians. In India, this initiative (also known as Free Basics) is based on a tie-up with one of the country's largest conglomerates, Anil Ambani's Reliance Communications. It proposes to deliver limited free Web content—38 websites, including access to Facebook, Wikipedia, Reliance Astrology, and Microsoft Bing—a move that has been critiqued by the civil society in India for compromising 'net neutrality' and favouring a monopoly provider, Reliance Communications.[62] The Indian government's active involvement in censoring access to the net—including banning sites involved in pornography and those advocating separatism (such as radical Sikh and Kashmiri organizations); threatening ISPs for hosting sites that are a security threat; and demanding access to encrypted online traffic linked to services such as Blackberry for surveillance and monitoring purposes—has also been critiqued by civil society, although key players such as Facebook and Google who are keen to capture the Indian market have, for the most part, acceded to the requests by the GoI to censor online traffic and sites.[63]

[61] Alawadhi, N. 2015. 'India Announces Open Source Policy: Big Win for FOSS', *The Economic Times*, 29 March, available at http://tech.economictimes. indiatimes.com/news/technology/india-announces-open-source-policy-big-win-for-foss/46738776; accessed on 10 September 2017.

[62] See Mishra, L. and S. Srinivasan. 2015. 'Facebook launches Internet.org in India', *The Hindu*, 11 February, available at http://www.thehindu.com/business/ Industry/facebook-launches-internetorg-in-india/article6879310.ece; accessed on 10 September 2017.

[63] Madon, Z. 2015. 'Facebook and Digital India: Bedfellows in online censorship?', *The Huffington Post*, 14 November, available at http://www.

India ranks among the top nations that demand online censorship from search engines such as Google and social networking sites such as Facebook. Aggressive censorship affects access and compromises the rights of citizens to a diversity of views and opinions. When this is couched within the subtext of Hindu nationalism, a key strength of India's democracy that is enshrined in its Constitution—that is, support for the diversity of cultures, expressions, and opinion—is at risk.

***

Algorithms such as EdgeRank, PageRank, and Epagogix enable prediction and ranking that are key to the sifting and sorting out of digital information, which in turn is key to the expansion of consumerism and the market. What makes algorithms intriguing is that for the most part, their sorting and ranking of personal information, that is then sold to the market, has not been a cause for any serious public concern; and that certainly has been the case in India. When the Snowden revelations first broke out, the newspapers that broke that story, including *The Guardian* and *The Washington Post*, did carry pieces on the State–corporate nexus and its impact on privacy. However, in the context of the persistent threat of terrorism, both imagined and real, the impact of such complicities has been ignored. Users tend to be pragmatic and as long as they can continue with their socialities online, they are not, on the whole, as concerned with how their personal data is being used. I would argue that one of the reasons for this state of affairs is that the aggressive marketing of personal information is not obvious; indeed, the warmth of a personal page that reflects the user's likes in terms of movie preference, products, and music offers a space where one can relax precisely because the threat levels are not obvious. However, as Siva Vaidhyanathan has argued, Google as both verb and noun affects us in a variety of ways:

> Googlization affects three large areas of human concern and conduct: 'us' (through Google's effects on our personal information, habits, opinion, and judgments); 'the world' (through the globalization of a strange kind of surveillance and what I'll call *infrastructural imperialism*); and 'knowledge'

huffingtonpost.in/zubin-madon/facebook-digital-india-st_b_8531394.html; accessed on 11 September 2017.

(through its effects on the use of the great bodies of knowledge accumulated in books, online databases, and the Web).[64]

The globalization of information is based on the globalization of algorithmic regulation. The collection of often-surreptitious primary data and metadata on citizens and consumers by the market and the State constitutes one of the key sources of power and control.

Chapter 3 explores the issue of software patents, an issue that is central to 'accumulation' in the knowledge economy.

---

[64] Vaidhyanathan, S. 2011. *The Googlization of Everything (And Why We Should Worry)*. Berkeley: University of California Press, p. 2.

3

Software Patent Manoeuvres

One of the major issues facing global information and communication today is the challenge from multinational companies and governments that are supportive of the expansions of protection for IP. The story of the global expansion of IP is tied to the Trade-Related Aspects of Intellectual Property Rights (TRIPS) Agreement that was negotiated at the end of the General Agreement on Trade and Tariffs (GATT) multilateral negotiations in 1994 and the establishment of the World Trade Organization (WTO) in 1996. The TRIPS, of course, became a cornerstone of WTO's trade negotiations, and all countries involved in the WTO had to abide by the key principles: national treatment, most favoured nation status, and reciprocity. While there was room for some exceptions, in general, TRIPS principles applied to all countries and importantly, reinforced and extended IP rights to cover areas previously outside of its purview. The emergence of the TRIPS agenda is clearly linked to the lobbying by governments, especially the US government, and key US MNCs to protect their competitive advantage in the leisure, high-technology, and information industries. From the mid-1970s, the Office of the United States Trade Representative (USTR) used a trade-related process known as Section 301 to globally monitor IP violations and to link this to trade sanctions. For the US multinationals, the key lobby was the Intellectual Property Committee (IPC) that was formed in 1986 to specifically integrate IP rights within the multilateral GATT negotiations. Peter Drahos, IP scholar, has commented on the membership of the IPC: 'The analogy with Hercules is apt for the membership of the IPC consisted of Bristol-Myers, Du Pont, FMC Corporation, General Electric, Hewlett-Packard, IBM, Johnson & Johnson, Merck, Monsanto, Pfizer, Rockwell International and Warner

Communication.'[1] I have written elsewhere on the politics of Section 301 as it was applied to the industrial and leisure sectors in India, particularly in relation to Hollywood imports.[2]

Given this historical pedigree of IP protection, the fact that it has become a critical aspect of the knowledge economy is not surprising. What is interesting from a geopolitical perspective is the many ways in which the USA in particular has used a variety of forums, from the WTO to WIPO, and a range of business lobbies inclusive of the Business Software Alliance and the Motion Picture Association of the USA along with multilateral, regional, and bilateral means to strengthen the remit of IP protection and pressure countries to create IP laws and policies that are in sync with the US and global IP laws. While such pressures have been contested, with the exception of the BRICS (Brazil, Russia, India, China, and South Africa) nations, a large percentage of countries in the Global South have had little choice but to establish IP policies supportive of the interests of transnational capital.

From information and communication perspectives, the politics and impact of copyright in relation to the global media and information industries have been the focus of numerous academic explorations. There have been studies from political, economic,[3] and cultural[4] perspectives and critical writings on the issue of cultural piracy that have attempted to deal with piracy from the bottom-up as it were.[5] Liberal scholars, mainly from the USA, have written extensively on the nature of contemporary

---

[1] Drahos, P. 1995. 'Global Property Rights in Information: The Story of TRIPS at the GATT', *Prometheus* 13(1): 12.

[2] Thomas, P.N. 2003. 'GATS and Trade in Audio-Visuals: Culture, Politics and Empire', *Economic and Political Weekly* 38(33): 3485–93.

[3] Bettig, R. 1996. *Copyrighting Culture: The Political Economy of Intellectual Property*. Colorado and Oxford: Westview Press; and Drahos, P. and J. Braithwaite. 2003. *Information Feudalism: Who Owns the Knowledge Economy?* New Delhi: Oxford University Press.

[4] Coombe, R. 1998. *The Cultural Life of Intellectual Properties: Authorship, Appropriation, and the Law.* Durham and London: Duke University Press.

[5] Karaganis, J. (ed.). 2011. *Media Piracy in Emerging Economies.* New York: Social Science Research Council; Liang, L. 2005. 'Porous Legalities and Avenues of Participation', *Sarai Reader: Bare Acts* 5: 6–17; and Sundaram, R. 2009. *Pirate Modernity: Delhi's Media Urbanism.* New York and London: Routledge.

copyright enclosures[6] and there are now a variety of studies on the issue of patenting in the life sciences and biotechnology.[7]

While copyright remains the key means used to create property across multimedia texts, the patenting of software has become a further means to protect software as property. However, and in spite of most governments around the world relenting on the issuance of software patents, it remains a contentious issue that is currently being fought in courts both in the developed and the developing world. Section 101 of the US Patent Act stipulates that patents are issued when a given invention passes the following tests: (i) that it is new; (ii) that it is useful; and (iii) that it is non-obvious; in other words, the test of novelty and inventiveness.[8] The patent system exists to provide an incentive to those who create and, more importantly, 'reveal' any given invention, thus ensuring that the idea gets added to the stock of ideas available in the public domain. Currently, there are three broad categories of patents: utility, design, and plant-based patents. While these look obvious enough, in reality patents can cover 'products, methods, apparatus, materials or processes that are new and

[6] Lessig, L. 1999. *Code and Other Laws of Cyberspace*. New York: Basic Books; Lessig, L. 2001. *The Future of Ideas: The Fate of the Commons in a Connected World*. New York: Random House; Lessig, L. 2004. *Free Culture: How Big Media Uses Technology and the Law to Lock Down Culture and Control Creativity*. New York: Penguin Press; and Vaidhyanathan, S. 2001. *Copyrights and Copywrongs: The Rise of Intellectual Property and How It Threatens Creativity*. New York and London: NYU Press.

[7] See Bowring, F. 2003. *Science, Seeds and Cyborgs: Biotechnology and the Appropriation of Life*. London and New York: Verso; Boyle, J. 1996. *Shamans, Software and Spleens: Law and the Construction of the Information Society*. Cambridge, MA, and London: Harvard University Press; Rajan, K.S. 2006. *Biocapital: The Constitution of Postgenomic Life*. Durham and London: Duke University Press; Rose, N. 2001. 'The Politics of Life Itself', *Theory, Culture & Society* 18(6): 1–30; Thacker, F. 2006. *The Global Genome: Biotechnology, Politics, and Culture*. Cambridge, MA, and London: MIT Press; and Waldby, C. and R. Mitchell. 2006. *Tissue Economics: Blood, Organs and Cell Lines in Late Capitalism*. Durham and London: Duke University Press.

[8] *Evaluating Subject Matter Eligibility Under 35 USC 101: August 2012 Update*, Office of Patent Legal Administration, United States Patent and Trademark Office, August 2012, available at http://www.uspto.gov/sites/default/files/patents/law/exam/101_training_aug2012.pdf; accessed on 13 March 2019.

sufficiently inventive'; and in biotechnology, it can include 'Agrichemistry, animal and human health products, biochemistry and protein chemistry, combinatorial and organic chemistry, gene therapies, genomics and genetic engineering, immunology and microbiology, cellular and molecular biology, plant biology, stem cell biology'; and in the pharmaceutical industries, it can include 'product formulation, medicinal chemistry, drug design, pharmacology'.[9]

By law, information on patents is public knowledge. Each patent is marked with a number and includes detailed information, including drawings. This information is published and is public precisely because the IP system, which in general includes copyrights and patents, is a contract given to IP holders for a period during which it is protected and cannot be exploited by anyone else except the IP holder. In other words, 'disclosure' is a fundamental aspect of the patent. The United States Patent and Trademark Office (USPTO) website, for example, includes full-text patent databases that are open for public searches.[10] The permissibility under the current patent system varies from country to country: for example, the patenting of genetic material remains controversial. In India, Section 3(i) under the Indian Patents Act, 1970, explicitly states that 'any process for the medicinal, surgical, curative, prophylactic, diagnostic therapeutic or other treatment of human beings or any process for a similar treatment of animals to render them free of disease or to increase their economic value or that of their products' is not patentable.[11] The Guidelines for Examination of Biotechnology Applications for Patent (1–23), prohibits 'plants and animals in whole or any part thereof other than microorganisms, but including seeds, varieties and species' from patentability.[12] The recent controversy related to the US-based company, Myriad Genetics—which owns patents to two genes, BRCA1 and

---

[9] Davies Collison Cave, IP, available at http://www.davies.com.au/content/47/techskills/life-sciences; accessed on 30 June 2017.

[10] USPTO, available at http://www.uspto.gov/blog/ebiz/; accessed on 19 June 2017.

[11] 'Section 3, Indian Patents Act, 1970', available at http://ipindia.nic.in/IPActs_Rules/updated_Version/sections/ps3.html; accessed on 9 June 2017.

[12] The Guidelines for Examination of Biotechnology Applications for Patent. 2013. Office of the Controller General of Patents, Designs and Trademarks, March, VIII Section 3(j): 4.

BRCA2, that are susceptible to mutation and can lead to ovarian or breast cancer, and thus controls the diagnostic testing of these genes—has raised the issue of gene monopolies and its impact on research, diagnosis, access, and the political economy of healthcare.[13]

Software Patents in the USA and Europe

The patenting of software is equally complex and contentious given that, at first glance, software is based on text and text traditionally falls under the realm of copyright law. However, the patenting of software has been a live issue in the USA, in particular from the early 1980s, and the government's position has changed from a no-patents attitude to an acceptance of patents and in recent years, to a renewed scepticism on the value of software patenting. Part of the reason for these changes in official attitudes are the pressures that are part of the peculiar economic environment in the USA, including the power of software lobbies and a government that is keen to protect the competitive advantage of American industry through creative interpretation and invocation of all the means that are available in the IP toolbox. In fact, software patents have been granted in the USA from the late 1970s and as pointed by Bessen and Hunt, between 1976 and 1999, at least 130,650 patents were granted (p. 8).[14]

> These industries, including the computer, electrical equipment and instruments industries, are also found to account for a major share of the growth in patenting in recent years. … Some researchers have suggested that firms in these industries may patent heavily in order to obtain strategic advantages, including advantages in negotiations, cross-licensing, blocking competitors, and preventing suits. … In principle, strategic patenting can arise whenever individual products involve many patentable inventions and the cost of obtaining patents is sufficiently low. … Firms may acquire large numbers of patents so that even if they have an unsuccessful product, they can hold up rivals, threatening litigation. Innovative firms may acquire 'defensive' patent

[13] Servick, K. 2015. 'End of the Road for Myriad Gene Patent Fight', *Science Insider*, 28 January, available at http://news.sciencemag.org/biology/2015/01/end-road-myriad-gene-patent-fight; accessed on 7 June 2017.

[14] Bessen, J. and R.M. Hunt. 2004. 'An Empirical Look at Software Patents', Working Paper No. 03-17/R., pp. 1–53, available at http://www.researchoninnovation.org/swpat.pdf; accessed on 4 June 2017.

portfolios to make a credible counter-threat. The outcome may involve the cross-licensing of whole portfolios, where firms agree not to sue each other and those firms with weaker portfolios pay royalties.[15]

Timothy Lee has observed in an article in *The Washington Post* that it is also a consequence of patent-law cases being removed from the purview of generalist judges in the Supreme Court in the 1980s and being moved to the Federal Circuit Appeals Court, whose judges are a lot more familiar and sympathetic towards patent lawyers.[16] In the light of the perceived laxity with which software patents had been issued, the Supreme Court overruled Federal Circuit patents on human genes in 2013, and on 31 March 2014, it discussed the case of a software patent after a hiatus of 33 years.[17] In contrast, the European Union (EU) rejected the Directive on the Patentability of Computer-Implemented Inventions (CII) in 2006 that was proposed by the lobby group, the European Information and Communication Technology Industry Association (EICTIA), which included firms such as Nokia, Microsoft, Siemens, and Alcatel. Article 2 of the Proposal for a Directive of the European Parliament and of the Council on the Patentability of Computer-Implemented Inventions defined it in the following terms: '(a) "computer-implemented invention" means any invention the performance of which involves the use of a computer, computer network or other programmable apparatus and having one or more *prima facie* novel features which are realised wholly or partly by means of a computer program or computer programs.'[18] The EICTIA was opposed by a loose grouping of NGOs committed to the open source and free software movements.[19] However, in spite of such

---

[15] Bessen and Hunt, 'An Empirical Look at Software Patents', p. 8.

[16] Lee, T.M. 2014. 'Will the Supreme Court Save us from Software Patents?', *The Washington Post*, 26 February, available at http://www.washingtonpost.com/blogs/the-switch/wp/2014/02/26/will-the-supreme-court-save-us-from-software-patents/; accessed on 4 June 2017.

[17] Lee, 'Will the Supreme Court Save Us from Software Patents?'

[18] 'Proposal for a Directive of the European Parliament and of the Council on the Patentability of Computer-Implemented Inventions', 2002/C 151 E/05, 20 February 2002, available at http://eur-lex.europa.eu/LexUriServ/LexUriServ.do?uri=COM:2002:0092:FIN:EN:PDF; accessed on 11 June 2017.

[19] Briendl, Y. 2010. 'Internet-Based Protest in European Policy-Making: The Case of Digital Activism', *International Journal of E-Politics* 1(1): 57–72, available

rulings, the European Patent Office has been involved in issuing patents for 'technical' inventions in software, although this stands in contrast to national jurisprudence, such as in the UK, where the Patent Act, 1977 excludes computer programmes from patentability, even though both Germany and France have issued patents for specific technical inventions that provide solutions to technical problems. Ballardini has highlighted the contrast between the straightforward patenting for traditional, tangible inventions and their functionalities and computers: 'In contrast, a computer-related invention rarely possesses any geometrical representation, since its components have no physical implementation and its result is intangible. It is this [sic] far harder to evaluate the concrete and "technical" applicability of a computer-related invention, and more interpretation is required in order to assess its patentability.'[20]

Software as Text and 'Behaviour'

There is another more fundamental reason for the confusions related to IP in software and the instruments to be used to protect this IP. All software consists of a set of instructions—in other text that is encapsulated in its source code. As such, the source code is eligible for copyright protection precisely because it is text, be it in the language of bits and bytes. However, the programmes and instructions that are a part of the source code need to be translated into its 'object code' in order for the programme to deliver what it is supposed to deliver. The object code delivers the functionalities of every software programme. This interaction between two types of code makes software different from ordinary products that can either be categorized under literary expression and therefore be placed under copyright or be categorized as an invention and placed under the patent system. This complexity is best explained in an article by Samuelson et al., where they clarify the difference between the software programme as 'text' but also critically as 'behaviour'—in other words, what it executes in

at http://www.reseaudel.fr/wp-content/uploads/2014/01/pdf_IJEP_ybreindl_final.pdf; accessed on 7 June 2017.

[20] Ballardini, R.M. 2008. 'Software Patents in Europe: The Technical Requirement Dilemma', *Journal of Intellectual Property Law & Practice* 3(9): 565.

terms of its functions.[21] For IP purposes, text and behaviour can be seen to be independent of each other and therefore needing specific types of protection:

> While conceiving of programs as texts is not incorrect, it is seriously incomplete. A crucially important characteristic of programs is that they behave; programs exist to make computers perform tasks. … Copyright law is mismatched to software, in part, because it does not focus on the principal source of value in a program (its useful behavior) … program text and behavior are largely independent … Copyright law does not protect the behavior of physical machines.[22]

Samuelson et al.,[23] along with Diver,[24] have made a case for a sui generis IP system for software, although arguably there is a case to be made that less IP rather than more is supportive of innovation, especially as it relates to Free and Open Source Software (FOSS) based expressions of software and the development of software in the developing world. Also, there are major questions related to the efficiency of the present system of software patents and to whether or not it has contributed to cumulative innovation. The fact that there continue to be divergent judgements in courts around the world on software patents does suggest that it remains difficult to judge the specificity and quality of software as an entity, given that it is different from conventional goods and services. In very significant ways, the very fact that the knowledge economy has become ubiquitous and has begun to impact on the quality of lives that people lead does suggest more of a role for FOSS-based solutions, as is the case with public sector

---

[21] Samuelson, P., R. Davis, M.D. Kapor, and J.H. Reichman. 1994. 'A Manifesto Concerning the Legal Protection of Computer Programs', *Columbia Law Review* 94(8): 2308–431.

[22] Samuelson et al. 'A Manifesto Concerning the Legal Protection of Computer Programs', pp. 2316, 2350. Also see Gonzalez, A.G. 2006. 'The Software Patent Debate', *Journal of Intellectual Property Law & Practice* 1(3): 196–206.

[23] Samuelson et al. 'A Manifesto Concerning the Legal Protection of Computer Programs'.

[24] Diver, L. 2008. 'Would the Current Ambiguities within the Legal Protection of Software be Solved by the Creation of a Sui Generis Property Rights for Computer Programs?', *Journal of Intellectual Law & Practice* 3(2): 125–38.

software in India. Given the thousands of software patents that are issued on a yearly basis, the possibility for numerous heterogeneities in the areas of interoperability and systems is a real probability. This reality becomes an issue in the context of government investments in e-government, when the choice is between investing in a proprietary system or navigating this heterogeneity. Bountouri et al. highlight the various levels that can be affected and that are a consequence of differing standards/protocols and heterogeneity in software applications:

> syntax—heterogeneities caused by the differences between protocols, encodings and languages used by information sources (i.e. query languages, data formats etc.);

> schema—heterogeneities coming from the implementation of different data models, data structures and schemas;

> semantic—heterogeneities produced by semantic conflicts arising from the fact that the meaning of the data can be expressed in different ways and with different interpretations; and

> system—heterogeneities arising from different hardware platforms, operating systems and networking protocols.[25]

Excessive patenting, or for that matter extensions of copyright, can have a limiting effect on the development of software for the public good.

This consequence of excessive IP protection impacts not only extended access projects supported by the public sector, but also affects innovation in FOSS and acts as an unnecessary curb on software innovation in general.

Innovation in the Information Commons

One of the major challenges in software innovation that is yet to be given serious thought relates to the very culture of software innovation in the FOSS sector. It is common knowledge that the strength of FOSS is a culture of collaboration and commitment to 'openness' as a fundamental

---

[25] Bountouri, L., C. Papatheodorou, V. Soulikias, and M. Stratis. 2009. 'Metadata Interoperability in Public Sector Information', *Journal of Information Science* 35(2): 205.

moral and social value. In the context of a resolutely neo-liberal global economy, sharing and collaboration are only grudgingly accepted by the industry given that innovation here is not based on competitive advantage or, for that matter, on the profit incentive. Recent interest in the value of the 'informational commons' has led to some interesting discussions on the specific nature of collaborative labour online, with Graham Murdock describing it as a reflection of a 'gift economy'. Murdock has argued that there are three levels to digital gifting—the sharing of self-produced material, cooperation in expanding opportunities for digital gifting, and collaboration in creating cultural products that are freely shared.[26] This reflects the growing interest in the concept of the informational commons.

Benkler describes the informational commons in the following terms. '"The commons" refers to institutional devices that entail government abstention from designating anyone as having primary decision-making power over use of a resource. A commons-based information policy relies on the observation that some resources that serve as inputs for information production and exchange have economic or technological characteristics that make them susceptible to be allocated without requiring that any single organization, regulatory agency or property owner, clear conflicting uses of the resource'. Benkler goes on to describe the informational commons in terms of the processes and resources that together permit information to be universally available to all people such as the electromagnetic spectrum that currently is in danger of becoming privatised.[27]

At the heart of the debate on the commons is a strong belief that access to the commons ought to be a fundamental human right, be it offline or online. The steady erosion of the commons offline, it is argued, has been matched by declines online, with exclusive forms of privatized access replacing universal access to the commons. Bollier has argued that the language of the commons stands in opposition to the language typically

---

[26] Murdock, G. 2011. 'Political Economies as Moral Economies: Commodities, Gifts and Public Goods', in J. Wasko, G. Murdock, and H. Sousa (eds), *Handbook of Political Economy of Communications*. Hoboken, NJ: Wiley-Blackwell, Hoboken, NJ, p. 25.

[27] Benkler, Y. 1998. 'The Commons as a Neglected Factor of Information Policy', available at http://www.benkler.org/commons.pdf, accessed on 21 November 2010, pp. 2, 11–21.

used to describe private 'property, contracts and markets'. The language of the commons, in contrast, is imbued with norms and rules and larger cultural rights that facilitate 'a richer, more qualitative and humanistic set of criteria that are not easily measured, such as moral legitimacy, social consensus and equity, transparency in decision making, and ecological sustainability, among other concerns'.[28]

However, this comeback of the commons needs to be grounded and as Vemuri has observed in the context of the FOSS movement, there is 'need for a new framework to understand innovation that is fuelled by enhancing one's reputation, recognition, understanding of programming, sense of belonging to group',[29] and dare I say, the cultures of collaboration, volunteerism, and mutual exchange. It is plain to see that software is built on the prior history of incremental innovations and as such, the nature of innovation in software requires another approach to its IP.

Software patenting remains an inexact science, meaning that it remains open to different interpretations, legal judgements, and therefore to on-the-ground consequences. The grant of a patent to Amazon.com's one-click order method (US Patent No. 5,960,411) for such an ordinary business function highlights the absurdity of the current IP system. Countries such as India have been pressured by both local and international software vendors to relax their position on software patents. In the case of India, negotiating such demands have not been straightforward and this has led to uncertainty and to a situation where some large software vendors have been granted patents and others not. In the USA, there has been a belated recognition of the need to curb the granting of patents for what are frivolous 'inventions' related to software. The Leahy–Smith America Invents Act, 2011, for example, includes provisions to challenge a patent post its granting and within a nine-month period.[30] This was preceded by a ruling in the *Bilski et al. v. Kappos* case, where a business

[28] Bollier, D. 2007. 'The Growth of the Commons Paradigm', in C. Hess and E. Ostrom (eds), *Understanding Knowledge as a Commons: From Theory to Practice*. Cambridge, MA: MIT Press, p. 29.

[29] Vemuri, V.K. 2004. 'Will the Open Source Movement Survive a Litigious Society?', *Electronic Markets* 14(2): 115.

[30] 'Leahy–Smith America Invents Act'. 2011, available at http://www. uspto.gov/sites/default/files/aia_implementation/20110916-pub-l112-29.pdf; accessed on 3 June 2017.

method was not granted a patent. The judgement specifically mentions that: '3. The applications process fails to produce a useful concrete and tangible result making it patent eligible.'[31] Sherly Abraham, in an article on software patents in the USA, has observed: 'The Bilski decision provides uncertain guidelines for the future of software patents. The Court has showed inconsistency in past decisions pertaining to software and business methods patents.'[32] The article also mentions reforms at the USPTO aimed at improving patent investigation and the need for a more transparent and robust patent review programme and search process. The need to reform the patent search process remains a key issue given that patents are granted in exchange for their public disclosure. When the information on any given software patent is obscure, badly written, hard to search for because of a lack of adequate keywords to help with the search, or is not properly 'marked', then it becomes difficult to assess whether or not one is infringing on a patent. According to Lindholm, it is a transparent disclosure regime that is a fundamental acid test for software patenting.[33] Without such a regime in place, software patenting lacks the accountability that other tangible 'inventions' simply have to conform to: 'All of the economic benefits depend to some degree on patent claims and technical disclosures being practically available to the public: invention (claims), designing around (claims and technical disclosure), disclosure (technical disclosure), and commercialization (claims). The problem must be solved if there is to be any pretense of an economic rationale for the software patenting regime.'[34]

The issue related to the need for a straightforward process for, and communication of, disclosure is a consequence of the more than 20,000 software patents that have been issued in the US alone.[35] These thousands

---

[31] *Bilsky et al. v. Kappos.* 2010. Supreme Court of the United States, available at http://www.supremecourt.gov/opinions/09pdf/08-964.pdf; accessed on 8 June 2017.

[32] Abraham, S.E. 2009. 'Software Patents in the United States: A Balanced Approach', *Computer Law & Security Review* 25(6): 558.

[33] Lindholm, S. 2005. 'Marking the Software Patent Beast', *Stanford Journal of Law, Business & Finance* 10(2): 82–128.

[34] Lindholm, 'Marking the Software Patent Beast', p. 96.

[35] Bessen, J. and R.M. Hunt, 'An Empirical Look at Software Patents', Paper presented at Conference on IPR, Innovation and Economic Performance, OECD,

of patents have led to what critics have referred to as 'patent thickets' that have become difficult to negotiate for companies and individuals who are keen to avoid the threat of patent infringement. These thickets curb innovation given that it does not facilitate any meaningful understanding of the 'prior art' that exists in software patents. These thickets are a particular obstacle to innovation in the FOSS movement that is dependent on the availability of and access to source code held by patent offices and those that are available in the public domain. Gonzalez highlights the nature of these thickets and their consequences:

> There are two types of patent thickets. The first one is a single technological innovation that may be protected by several patent holders. This situation would require anyone interested in developing software in that area to obtain separate licenses from numerous owners. The second type of thicket occurs when a product is covered by a large number of patents, not just one. Patent thickets increase the cost of innovation, they create inefficiency through the creation of complex cross-licensing relations between companies and they may even stop newcomers entering the market if they fail to penetrate the thicket.[36]

The Case of Software Patents in India

The move by the Indian government to include software patenting within the purview of The Patents (Amended) Act, 25 June 2002, certainly has been of concern to IP activists in India and abroad.[37] Their inability to form a critical mass and advocate from a position of strength stands in contrast to both the position taken by the World Trade Organisation and software multinationals that support the extension and expansion of software patents. Furthermore, the presence of the world's leading software multinationals in India and the ambitions of the leading Indian software export companies to protect their IP in an increasingly competitive software context have also contributed to the opening up of IP possibilities in software.

28–9 August 2003, available at http://www.oecd.org/sti/sci-tech/11742356.pdf; accessed on 17 June 2017.

[36] Gonzalez, 'The Software Patent Debate', p. 204.

[37] 'The Patents (Amendment) Act, 2002', available at http://ipindia.nic.in/ipr/patent/patentg.pdf, accessed on 8 September 2010.

While it is a fact that the Indian Patents Act does specify that some inventions including 'a mathematical or business method or a computer program *per se* or algorithms',[38] *cannot* be patented, this formulation does lend itself to the suggestion that a programme that is integrated into a device *can* be patented. The term 'per se' lends itself to multiple interpretations. While a software program 'per se' may not be patented it does suggest that any proof of a 'technical effect' in software or an algorithm can become the basis for its patenting. While Patent Amendments Ordinance, 2004, put forward the view under Section 3 that patents could include '(k) a computer programme per se other than its technical application to industry or a combination with hardware; (ka) a mathematical method or a business method or algorithms',[39] this Ordinance was rejected by Parliament in 2005 and, as a consequence, the expansion of the definition as contained in the Ordinance, was not included in the Patent (Amendment) Act, 2005. The GoI has rejected software patent applications. In 2011, the Intellectual Property Appellate Board applied Section 3(k) to an application from Yahoo for a 'method of operating a computer network search apparatus', but rejected it for being an unpatentable business method.[40]

However, support for software patents has appeared in other texts. The Draft Manual of Patent Practice and Procedure issued by the Patent Office makes a clear case in favour of software patents. This is highlighted by Section 4.11.10: 'A mathematical method is one which is carried out on numbers and provides a result in numerical form (the mathematical method or algorithm therefore being merely an abstract concept prescribing how to operate on the numbers) and not patentable.

---

[38] See, The Patents Act. 1970, available at http://www.ipindia.nic.in/writereaddata/Portal/IPOAct/1_113_1_The_Patents_Act_1970_-_Updated_till_23_June_2017.pdf, p. 3, 8, Section 3K; accessed on 14 March 2019.

[39] 'The Patents (Amendment) Ordinance, 2004', available at http://lawmin.nic.in/Patents%20Amendment%20Ordinance%202004.pdf, accessed on 10 September 2010.

[40] Intellectual Property Appellate Board. 2011. 'Yahoo Inc. vs Assist. Controller of Patents and Designs & Rediff.com India Ltd.', 8 December, available at http://www.ipabindia.in/Pdfs/Order-222-11-OA-22-10-PT-CH.pdf; accessed on 8 June 2017.

However, its application may well be patentable, for example, in *Vicom/ Computer-related invention* [1987] 1 OJEPO 14 (T208/84) the invention concerned a mathematical method for manipulating data representing an image, leading to an enhanced digital image'.[41] Arguably, such amendments can be viewed as attempts by the Government of India to harmonise its IP laws with international laws such as the World Trade Organisation administered Trade Related Aspects of Intellectual Property Rights (TRIPS). The *Manual of Patent Office Practice and Procedure* likewise does not include wording supportive of software patents.[42]

Under 'Chapter XVI: Working of Patents, Compulsory Licences and Revocations', the 2002 Act does lay out the terms of the patent—the 'general principles applicable to the working of patented inventions':[43]

(c) that the protection and enforcement of patent rights contribute to the promotion of technological innovation and to the transfer and dissemination of technology, to the mutual advantage of producers and users of technological knowledge and in a manner conducive to social and economic welfare and to a balance of rights and obligations;[44]

Since software is unlikely to be the cause for any national emergency, the government will not be predisposed to revoking software patents

[41] *Draft Manual of Patent Practice and Procedure*. 2008. India: The Patent Office, 2008, p. 74, available at http://ipindia.nic.in/ipr/patent/DraftPatent_Manual_2008.pdf, accessed on 8 September 2010.

[42] *Manual of Patent Office Practice and Procedure* (as modified on 22 March 2011). The Office of the Controller General of Patents, Designs and Trademarks, Mumbai, 2011p. 91 onwards, available at <http://www.ipindia.nic.in/ipr/patent/manual/HTML%20AND%20PDF/Manual%20of%20Patent%20Office%20Practice%20and%20Procedure%20-%20pdf/Manual%20of%20Patent%20Office%20Practice%20and%20Procedure.pdf>.

[43] The Patents Act. 1970. Section 83. General Principles Applicable to Working of Patented Inventions, Clause C, available at http://ipindia.nic.in/writereaddata/Portal/ev/sections/ps83.html; accessed on 13 March 2019.

[44] 'Clause C is the Exact Replica of Article 7 in the Uruguay Round Agreement: TRIPS, Part 1: General Provisions and Basic Principles, WTO. Available at https://www.wto.org/english/docs_e/legal_e/27-trips_03_e.htm; accessed on 13 March 2019.

unlike patents for pharmaceutical products.[45] However, the granting of software patents could well impact on the government's plans to invest in the creation of indigenous software to be used in e-governance and other public sector-related projects. Additionally, software patents can curb innovation by small- and medium-sized companies in India, who do not have the resources to negotiate with patent holders and will not be able to strengthen their own indigenous capacities in software development.

Grosche has convincingly argued that

> Copyright with some adaptations to software has proven to be a workable and internationally accepted forms of providing incentives for innovation in the software sector (rather than in patent litigation). Based on algorithms and abstract ideas of which it has always been recognised that they cannot be monopolised in a reasonable way, software is a phenomenon too different to allow for further "muddling through" by imposing the complex and expensive patent system onto a new field in which it is either opposed, ignored—or abused, and where both U.S. and European courts have tried in vain to reach legally consistent, let alone convincing solutions.[46]

Grosche's arguments need to be viewed in the light of multiple litigations over infringements in courts in the EU, the USA and elsewhere and, in response, attempts by the EU to establish a Uniform Patent Legal System, streamline patent processes and reduce litigation costs.[47] The End Patents Coalition has calculated that an average of 55 software litigations are filed each week, that it costs on average US$4 million per litigation and that close to US$11.4 billion is wasted on software litigation each year.[48]

[45] *Guidelines for the Examination of Computer Related Inventions (CRIs)*, Indian Patent Office, October 2013, available at: http://www.ipindia.nic.in/iponew/draft_Guidelines_CRIs_28June2013.pdf.

[46] Grosche, A. 2006. 'Software Patents—Boon or Bane for Europe?', *International Journal of Law and Information Technology* 14(3): 308.

[47] See Coyle, P. 2012. 'Uniform Patent Litigation in the European Union: An Analysis of the Viability of Recent Proposals Aimed at Unifying the European Patent Litigation System. *Washington University Global Studies Law Review* 11 (1): 1–23.

[48] Asay, M. 2008. '$11.4 Billion Wasted on Software Patent Litigation … and Counting', *The Open Road*, available at http://news.cnet.com/8301-13505_3-9882152-16.html, accessed on 23 September 2010.

The issue of software patenting remains a live issue in India, where in spite of Court decisions, a small amount of software patents have been issued. Chingale and Rao have observed that 'in the year 2009–10, a total of 1195 patents were granted against the 892 in 2010–11, 564 in 2011–12 and 510 in 2012–13'.[49] However, and as per this article, these patents have been granted in the 'field of information technology' and could include more than just software patents. In October 2013, the Indian Patent Office organized a stakeholder meeting where the document *Guidelines for the Examination of Computer Related Inventions (CRIs)* was released.[50] A report from the Software Freedom Law Centre (SFLC) on this meeting indicates that representatives from MNCs and their patent lawyers and patent agents tried to reignite discussions on software patents.[51] Software patents are banned under the 2016 *Guidelines on Computer Related Inventions (CRIs)*, although there is pressure on the government to ease these restrictions. Anju Srivas, writing in *The Wire*, has observed that:

> Multinational software companies are naturally displeased with this turnaround by the Indian patent office. In a statement, BSA (also known as the 'Software Alliance')—an industry lobby group whose members include Apple, Microsoft, Dell and IBM—has noted its displeasure, saying that 'it will continue to work with the government to ensure that software inventions continue to be eligible for patent protection.'[52]

---

[49] See Chingale, R. and S.D. Rao. 2015. 'Software Patent in India: A Comparative Judicial and Empirical Overview', *Journal of Intellectual Property Rights* 20: 217.

[50] Guidelines for Examination of Computer-Related Inventions (CRIs). 2017. Officer of the Controller of Patents, Designs and Trademarks, 2015, available at http://www.ipindia.nic.in/writereaddata/Portal/Images/pdf/Revised__Guidelines_for_Examination_of_Computer-Related_Inventions_CRI__.pdf; accessed on 14 March 2019.

[51] 'Software Patents: Putting the Genie Back into the Bottle', SFLC, 16 October 2013, available at http://sflc.in/software-patents-putting-the-genie-back-in-the-bottle/; accessed on 4 June 2017.

[52] Srivas, A., 'The Long, Drawn-Out Fight to Regulate Software Patents in India', *The Wire*, 24 February 2016, available at https://thewire.in/22539/the-long-drawn-out-fight-to-regulate-software-patents-in-india/; accessed on 11 June 2017.

In spite of the availability of guidelines, patent offices in India have been inconsistent in their rulings and have issued patents for MNCs such as Google and Facebook, but not for Indian companies. Devika Agrawal has highlighted the fact that 'Google was granted a patent on an invention titled, "phrase identification in an information retrieval system"'. In this case too, 'Google argued that its invention is not an algorithm or a computer program *per se*, "but provides a technical solution to a technical problem of how to automatically identify phrases in a document collection".'[53] The fact remains that despite provisions in the Patents Act, software patents have been issued. There are many ongoing efforts by both apex bodies in India and MNC advocacy groups, such as the Business Software Alliance, to make software patents legal.

Activist scholars in India, namely those associated with the Alternative Law Forum and the Centre for Internet & Society, both based in Bengaluru, along with members who belong to the FOSS community have made a strong case to curb the granting of software patents. Pranesh Prakash from the Centre for Internet and Society has observed that 'the patenting of software helps three categories of people: (1) those large corporations that already have a large number of software patents; (2) those corporations that do not create software, but only trade in patents/ sue on the basis of paatents ("patent trolls"); (3) patent lawyers.'[54] Liang, Sethi, and Iyengar make the point that software patents are a stronger method of protection than copyright, 'because the protection extends to the level of the idea embodied by the software and injuncts ancillary uses of an invention as well.'[55] The Free Software Movement of India has argued that 'software patents kills innovation' and that the *Guidelines*

[53] Agrawal, D. 2017. 'Software Patents: Prohibited under Indian law but Granted in Spirit', *Tech2*, 15 May, available at http://www.firstpost.com/tech/ news-analysis/software-patents-prohibited-under-indian-law-but-granted-in-spirit-3702725.html; accessed on 14 March 2019.

[54] Prakash, P. 2010. 'Arguments against Software Patents in India', CIS, available at http://www.cis-india.org/advocacy/ipr/blog/arguments-against-software-patents, accessed on 24 September 2010.

[55] Liang, L., A. Sethi, and P. Iyengar. n.d. 'Briefing Note on the Impact of Software Patents on the Software Industry in India', Sarai.net, p. 4, available at http://www.sarai.net/research/knowledge-culture/critical-public-legal-resources/ whysoftwarepatentsareharmful.pdf, accessed on 24 September 2010.

for the Examination of Computer-Related Innovations go against the spirit of the Patents Act, 1970.[56] Despite the recall of this guideline in 2015 and the issuance of a new guideline in 2016 for software patents in India, there is a lack of clarity since multiple agencies are involved in awarding patents. Despite the 2016 Guideline that awards software patents only to innovations that fulfil the 'novel hardware requirement', patents have been awarded to a select group of large vendors including Google, Facebook, and Apple by patent offices in India.[57]

Liang, Sethi, and Iyengar have also argued that the flexible, public interest-based nature of copyright protection in India has served the Indian industry well. They, however, make the point that these gains can be offset by the government granting patent protection. Such protections can interfere with and restrict 'interoperability'—a major concern in the context of the Indian government's vast investments in e-governance and interest in the development of public sector software.[58] Free software activists in India have certainly welcomed the position taken by the DeitY that 'makes it mandatory for all software applications and services of the government [to] be built using open source software, so that projects under Digital India "ensure efficiency, transparency and reliability of such services at affordable costs."'[59] All tenders for software by government departments now have to compare capability, strategic control, scalability, security, lifetime costs, and support requirements between open source and

[56] Free Software Movement of India: Software Patents Kills Innovation. n.d., available at https://fsmi.in/content/software-patents-kills-innovation; accessed on 14 March 2019.

[57] See Agarwal, 'Software Patents: Prohibited Under Indian Law but Granted in Spirit'. *Tech2*, 15 May, available at https://www.firstpost.com/tech/news-analysis/software-patents-prohibited-under-indian-law-but-granted-in-spirit-3702725.html; accessed on 14 March 2019.

[58] Liang, L., A. Sethi, and P. Iyengar. n.d. 'ALFs Note before 2005 Amendment: Briefing Note on the Impact of Software Patents on the Software industry in India', *Centre for Internet & Society*, available at https://cis-india.org/openness/publications/software-patents/alfs-note-before-2005-amendment; accessed on 14 March 2019.

[59] Alawadhi, N. 2015. 'India Double Downs on Use of Open Source Software', *The Economic Times*, 29 March, available at http://economictimes.indiatimes.com/tech/internet/india-doubles-down-on-use-of-open-source-software/articleshow/46738604.cms; accessed on 11 May 2017.

proprietary software. The open source software-based Bharat Operation Systems Solutions (BOSS), developed by NRCFOSS and maintained by the Chennai-based arm of C-DAC, is currently distributed in 19 Indian languages, and government departments have been instructed to adopt BOSS-based platforms that are ideal for localized solutions. Similarly, the Open Technology Centre, also based in Chennai, is involved in exploring open standards for e-governance and open source software-based solutions. While there are major savings in costs related to the adoption of open source software solutions, the fact that it can be customized by end users, that it bypasses vendor lock-ins and strengthens government's strategic control over software deployments, and also increases possibilities for scalability, have been key drivers in the government's adoption of this position. It also reflects a gradual turn towards mixed solutions in public software that have been adopted by many countries, both developed and developing. The government's position on open source software has led to investments by international open source companies, such as Cyanogen, and local companies, such as WIPRO, that see advantages in both adopting the patent route and investing in core technology initiatives based on open source software. However, given Cyanogen's close relationship with Microsoft, it is difficult to assess their larger game plan in India. Rather predictably, the response from apex bodies such as National Association of Software and Services Companies (NASSCOM) and proprietary firms such as Microsoft has been critical of the government's position on open source software. They have argued that the government should be 'technology neutral' and that the adoption of open source software should not become a mandatory requirement. However, the government's major IT initiatives, including the $18 billion Digital India project and its associated $11 billion National Optical Fibre Network initiative that is aimed at connecting 250,000 village clusters to the Internet,[60] require flexibilities in software that will not be served by the granting of software patents. At the state level, as opposed to the federal level, the transition to FOSS-based solutions continues to gain momentum. In the Indian state of Kerala—a state that has engaged with FOSS from the year 2001—its

---

[60] Toness, B.V. 2015. 'Narendra Modi Plans Rs 70,000 Crore Outlay to Bring Rural Poor Online', *Livemint*, 19 May, available at http://www.livemint.com/Politics/Led8pVLTsK3sCSW7fBKC4K/Narendra-Modi-plans-Rs70000-crore-outlay-to-bring-rural-poo.html.

legislative assembly decided in 2014 that all its business will be carried out on a free software platform.[61] In that same year, in the neighbouring state of Tamil Nadu, the withdrawal of support for Windows XP acted as a catalyst for government departments to transition to BOSS-based platforms and solutions.[62]

In light of the recent developments in India on open source software policy, it can be expected that the process of granting software patents will not be straightforward and will depend on whether or not the originality and inventiveness of software patent submissions will have a negative impact on the government's plans to invest in making software 'public.' While the Indian government may not go down the path taken by the Government of New Zealand that approved the Patents Bill in October 2013, which includes a clause under Other Exclusions 3(A) that 'A computer program is not a patentable invention,'[63] based on the evidence available, we can certainly expect it to embrace flexible policies that result from domestic compulsions as much as by reason of realpolitik.

[61] See Kurian, V. 2014. 'Kerala Legislature Announces Smooth Transition to Free Software', *Business Line*, 18 July, available at http://www.thehindu businessline.com/news/states/kerala-legislature-announces-smooth-transition-to-free-software/article6224551.ece.

[62] Ravi Kumar, N., 'State Departments Asked to Switch over to Free Open Source Software', *The Hindu*, 18 March 2014, available at http://www.thehindu. com/todays-paper/tp-national/tp-tamilnadu/state-departments-asked-to-switch-over-to-free-open-source-software/article5798429.ece.

[63] 'Patents Bill, 2013, New Zealand', available at file:///Users/uqpthom4/ Downloads/Patents%20Bill.pdf.

Section II

The Sovereign/Ambivalent State?

4

Digital (Transgenic) Seed and Its Copy

This chapter explores the issue of transgenic seed as copy against the background of the agrarian crisis in India, which is, in some measure, the obverse of the 'success' of the IT economy. This disproportionate growth is reflected in the fact that high growth in the services sector has been offset by slow growth or no growth in the industrial and agricultural sectors. In one of the early articles on digital capitalism, Dan Schiller made a case for critical political economists of communications to theorize information in terms of its application across multiple sectors and not in terms of discrete technologies.[1] The expansive footprint of digital capitalism along with multiple convergences has resulted in extensive commodification of information across all economically productive sectors. In Schiller's words:

> The transition to information capitalism does not depend on or equate with a narrow section of the media-based products. It is co-extensive with a socio-economic metamorphosis of information across a great (and still undetermined) range. As commodity relations are imposed on previously overlooked spheres of production, new forms of genetic and biochemical information acquire an unanticipated equivalence with other, more familiar, genres. Agribusiness, pharmaceutical giants, energy and chemical corporations, and medical companies—all essentially concerned with diverse genetic and bio-chemical information streams—are in the midst of a continuing technological transformation of the means of information production that is

---

[1] Schiller, D. 2007. *How to Think About Information.* Urbana and Chicago: University of Illinois Press, p. 25.

every bit as relevant to our understanding of the parallel trend 'convergence' between television, computing and telecommunications.[2]

Close to 70 per cent of rural households in India are involved in agriculture as their means of livelihood.[3] While agricultural policy in the immediate post-Independence years was directed towards subsidizing agriculture along with limited success in the redistribution of land, the 'Green Revolution' in the 1970s that led to greater agricultural productivity began the process of capitalization of Indian agriculture. This was based on high capital and resource investments and led to dependence on inputs such as fertilizers, pesticides, and hybrid seeds. While the Green Revolution certainly helped India become self-sufficient in the production of essential grains, it also led to the gradual immiseration of large numbers of small farmers who could not afford to invest in new and capital-intensive agricultural practices and had no access to resources such as the regular supply of water for irrigation. Arguably, the Green Revolution strengthened food security at the expense of food sovereignty. The liberalization of agricultural policies in the 1990s, inclusive of the withdrawal of agricultural subsidies, the decontrol of fertilizers, along with a decline in access to credit, was a critical factor that contributed to the agrarian crisis.

One of the consequences of limited borrowing opportunities for farmers from banks has been an increase in the 'moneylender's share in total credit', which 'increased from 10.7 per cent in 1991 to 25.7 percent in 2003', and has led to a spate of suicides directly attributed to farmer indebtedness.[4] It has been estimated that close to 270,000 farmers have committed suicide over the last two decades, a reality that the Indian poverty journalist P. Sainath, the rural affairs editor with *The Hindu*, has relentlessly exposed.[5] According to Sainath, the farmers' suicide rate (FSR) was 16.3 per 100,000 farmers

[2] Schiller, *How to Think About Information*, p. 25.

[3] 'FAO in India: India at a Glance, Food and Agriculture Organisation of the United Nations', available at http://www.fao.org/india/fao-in-india/india-at-a-glance/en/; accessed on 25 March 2019.

[4] See Bhattacharayya, S., M. Abraham, and A. D'Costa. 2013. 'Political Economy of Agrarian Crisis and Slow Industrialisation in India', *Social Scientist* 41(11–12): 50.

[5] Sainath, P. 2013. 'Farmers' Suicide Rates Soar above the Rest', *The Hindu*, 18 May, available at http://www.thehindu.com/opinion/columns/sainath/farmers-suicide-rates-soar-above-the-rest/article4725101.ece; 22 July 2018.

in 2011. He has also observed that farmer suicides have been the highest in AP and Maharashtra—while AP is currently being transformed into a high-technology state, Maharashtra is the state with the most concentrated disparities in wealth: 'A farmer in Andhra Pradesh is three times more likely to commit suicide than anyone else in the country, excluding farmers. And twice as likely to do so when compared to non-farmers in his own State. The odds are not much better in Maharashtra, which remained the worst State for such suicides across a decade.'[6] Kathy Le Mons Walker's Marxian reading of this crisis highlights the intense pressures placed on the Indian government by the US government first, followed by the International Monetary Fund (IMF) and World Bank, to liberalize its trade policies and focus on export-led agricultural growth and policies supportive of such growth.[7] A variety of factors, including the withdrawal of State subsidies for agriculture, vagaries in the international pricing of cash crops, and a series of price decreases that were compounded by cash crop failures in the 1990s—in cotton, jute, coffee, tea, and pepper—have all contributed to the agrarian crisis in rural India.

The agrarian crisis in India has also been precipitated by the turn towards the life science industries for the supply of a number of inputs, including, most importantly, seed. While the Indian government's seed policy was firmly in the public sector in the post-Independence period, following certain legislations—such as the Industrial Licensing Policy (1987) that led to the de-licensing of seed, Policy on Seed Development (1988), and the Protection of Plant Varieties and Farmers' Rights Act (PPVFRA, 2001)—along with India's accession to the WTO's TRIPS, the privatization of the business of hybrid seed has grown exponentially. India's response to TRIPS was to establish a sui generis policy, that is, the PPVFRA that supports both farmers' and breeders' rights as per TRIPS obligations. However, the PPVFRA's support of both *farmers'* rights and the *market* is problematic to say the least.[8] While the Act does give

[6] Sainath, 'Farmers' Suicide Rates Soar Above the Rest'.

[7] Walker, K.L.M. 2009. 'Neoliberalism on the Ground in Rural India: Predatory Growth, Agrarian Crisis, Internal Colonization and the Resurgence of the Class Struggle', *Journal of Peasant Studies* 35(4): 557–620.

[8] See Sahai, S. n.d. 'India's Plant Variety Protection and Farmers' Rights Act', available at http://www.iprsonline.org/ictsd/docs/SahaiBridgesYear5N8 Oct2001.pdf; accessed on 11 July 2018.

ordinary farmers the right to save and sell seed, Dewan has observed that the real beneficiaries are registered breeders:

> The PVP law in India benefits the registered breeder to save, use, sow, re-sow, exchange and share or sell his new variety and the breeder who obtains registration of a new plant variety can stop any person who sells, exports, imports or produces such variety without her permission. He can also prevent the use, sale, export, import or production of any variety deceptively similar to the registered variety.[9]

The pressure to strengthen plant breeders' rights in India was, to a large measure, the outcome of several key studies by seed companies in the late 1980s, alongside an influential study by United States Agency for International Development (USAID), which recommended that the government create the right environment and conditions for MNC research and supportive legislation, including a 'Plant Variety Protection Act'. While the Indian government in its PPVFRA has included a provision for compulsory licensing (Sections 8[2e] and 51[1ii]) that can be invoked if the public does not have access to seed for protected varieties and/or propagating material, India's projected accession to the International Union for the Protection of New Varieties of Plants (the UPOV Convention, 1991) is bound to result in the further dilution of farmers' rights to seed and to the increase of breeders' rights.[10] Article 14(9) in UPOV, 1991 severely restricts the rights of farmers to propagate and/or reuse seed,[11] and instead supports the rights of seed companies in the multiplication of seed, its propagation, sale, export, import, and stocking for such purposes.

Just as successive copyright amendments in the USA and elsewhere, backed by WIPO legislations, have extended the life of copyright, so have

---

[9] Dewan, M. 2011. 'IPR Protection in Agriculture: An Overview', *Journal of Intellectual Property Rights* 16(2): 136.

[10] 'The Protection of Plant Varieties and Farmers' Rights Act, 2001', available at http://www.plantauthority.gov.in/pdf/PPV&FRAct2001.pdf.

[11] See Dhar, B. and S. Chaturvedi. 1998. 'Introducing Plant Breeders' Rights in India: A Critical Evaluation of the Proposed Legislation', *The Journal of World Intellectual Property* 1(2): 245–62. http://www.upov.int/en/publications/conventions/1991/w_up911_.htm#_14; accessed on 22 July 2018.

seed legislations, such as the UPOV, 1978 and its 1991 version, extended the control of plant breeders' rights over the production, distribution, and use of such seeds. Together with the TRIPS Agreement, these legislations have progressively granted plant breeders—a euphemism for large MNCs such as Monsanto—extraordinary rights to control hybrid seed and its terms of use. The 1991 Act extends the terms of protection for 20 years and requires a 25-year term for tree and vine varieties. However, its complete control and exclusive rights over the uses of seed, as articulated in Article 14, has been criticized by small farmer groups and a number of countries from Africa for its attempt to restrict the culture of sharing seed, and also to make seed an exclusive commodity whose rights to use are controlled by the company that sells the seed, very much like Microsoft licenses its software for use strictly by the individual buyer.

a) Subject to Articles 15 and 16, the following acts in respect of the propagating material of the protected variety shall require the authorization of the breeder:

   (i)   production or reproduction (multiplication),

   (ii) conditioning for the purpose of propagation,

   (iii) offering for sale,

   (iv) selling or other marketing,

   (v)   exporting,

   (vi) importing,

   (vii) stocking for any of the purposes mentioned in (i) to (vi), above.[12]

What makes Article 14 particularly problematic for peasant farmers throughout the world is that it not only restricts their customary 'farmer's privilege' but also gives the breeders the right to collect royalties to both harvested crops as well as processed products.[13] This constant and cumulative strengthening of the rights of plant breeders to the detriment of farmers has been a cause for concern in the USA where anti-trust enforcement has constantly been denied. Between 2010 and 2012, the United States Department of Agriculture (USDA) did explore rising

[12] 'UPOV 1991, Article 14', available at http://www.upov.int/en/publications/conventions/1991/w_up911_.htm#_14; accessed on 22 July 2018.

[13] Boehm, T. 2013. 'Farmer's Privilege and UPOV '91', *The Union Farmer Quarterly*, Spring, p. 17.

concentrations in the ownership and control of agriculture inputs, including control over transgenic seed, although it has been reported that the findings were not made public.[14] Among the issues that have been explored are: whether or not the impact of licensing restrictions on growers, seed companies, and rival biotechnology firms fall within the purview of patent law; or whether or not this is a reflection of an attempt to use existing power asymmetries in the control over transgenic seed to shape the field and control competition. Monsanto's stacking of traits in transgenic seeds, both 'intra-firm' and 'inter-firm', is detrimental to competition in the seed industry:

> The competitive success of inter-firm stacking, however, is limited by a number of factors. First, the presence of a dominant traits 'platform' serves as a barrier to entry or expansion to competing inter-firm stacks that do not contain Monsanto traits. Indeed, the 30% of the market that is open to inter-firm stacks not containing Monsanto traits is occupied mostly by collaborations between the same three firms—Dow, Bayer, and Syngenta. Second, inter-firm stacking that involves collaborating with a dominant firm is potentially limited by licensing conduct of the sort that has been the subject of antitrust counterclaims in patent infringement cases. This includes selective or discriminatory royalties and cross-licensing or out licensing requirements.[15]

The many patent infringement cases in court linked to transgenic seed are a mirror image of controversies over software patent cases, the impact of 'stacking', and the resulting patent thickets that frustrate innovation and creativity by competitors. Such thickets are characterized by a dense web of patents and overlaps that are difficult to distinguish precisely because patents have been awarded for incremental improvements (see Chapter 3). Needless to say, this again is an illustration of the productive power of couplings between biotechnology and information technologies

---

[14] See Fuglie, K., P. Heisy, J. King, and D. Schimmelpfennig 2012. 'Rising Concentrations in Agricultural Input Industries Influences New Farm Technologies', USDA, Economic Research Service, 3 December, available at http://www.ers.usda.gov/amber-waves/2012-december/rising-concentration-in-agricultural-input-industries-influences-new-technologies.aspx#.VnsqxaN--os; accessed on 2 July 2018.

[15] Moss, D.L. 2013. 'Competition, Intellectual Property Rights, and Transgenic Seed', *South Dakota Law Review* 58(3): 555.

in contexts characterized by a high incidence of concentrated ownership. Complex productive technologies in highly competitive environments have resulted in firms both shaping, some would say 'gaming', the very environments of IP and the boundaries of possibilities.

Intellectual property–based enclosures, particularly in agriculture, have led to further marginalization of farmers in India. There has been a relentless migration of farm labour to the cities, especially for employment in the construction and allied sectors, as farmers find it difficult to make a living out of farming given the increase in the price of inputs and fluctuating prices of agricultural products. Harvey makes a case for the fact that IP is another aspect of the new imperialism and the means by which ABD further marginalizes the poor:

> Wholly new mechanisms of accumulation by dispossession have also opened up. The emphasis upon intellectual property rights in the WTO negotiations (the so-called TRIPS agreement) points to ways in which the patenting and licensing of genetic materials, seed plasmas, and all manner of other products, can now be used against whole populations whose environmental management practices have played a crucial role in the development of those materials.[16]

The Seed Industry in India

In 2013, the seed industry in India was worth $2.7 billion in revenue.[17] This industry is made up of five types of firms: technology firms, trading companies, small-sized seed firms, medium-sized firms, and MNCs. The MNCs are highly integrated with interests in seed, agricultural biotechnology (agbiotech), and agrichemicals; have extensive research and development (R&D) capacities; and are based on vertical and horizontal integration of their upstream and downstream operations. The major firms include BASF, Bayer, CropScience, Dow Agrosciences, DuPont,

---

[16] Harvey, 2004. 'The "New" Imperialism: Accumulation by Dispossession', *Socialist Register* 40: 75.

[17] Chintada, V. 2014. 'Progression of Indian Seed Industry', Sathguru Management Consultants, available at http://aginnovation.org/malawi/workshop/Progression-of-Indian-Seed-Industry_Mr.%20Venugopal-Chintada.pdf; accessed on 11 July 2018.

Monsanto, and Syngenta.[18] One of the less-studied aspects of these seed firms is their labour practices. A study on child labour and transnational seed companies in the state of AP highlights the role of child labour, especially young girls, in the 'cross-pollination' of hybrid cotton seed in the 200+ companies involved in hybrid cotton seed production in India. Typically, MNCs license local seed companies to produce seed. These local companies employ close to 450,000 children in the production of seed, out of which 287,000 work in the state of AP. The report estimates that:

> in the year 2000–2001, five multinational seed companies i.e. Syngenta, Hindustan Lever, Advanta, Proagro and Mahyco–Monsanto accounted for nearly 21.6% (5350 acres out of 24783) of the total area under hybrid cottonseed production in Andhra Pradesh. The number of children employed in farms producing and supplying seed for these MNCs is estimated to be around 53500. Out of 53500 children, HLL accounted for 25,500, Syngenta 6,500, Mahyco–Monsanto 17,000, Advanta 3,000 and Proagro 2,000.[19]

The International Politics of Seed

Michael Peters explains the term bio-informational capitalism as a 'form of capitalism based on a self-organising and self-replicating code that harnesses both the results of the information and new biology revolutions and brings them together in a powerful alliance that enhances and strengthens or reinforces each other.'[20] What is fascinating with respect to the politics of 'trans-genic' seed is that the disruptive potential of the digital is now seen as a possible means for creating solutions that can be widely accessed and used by the world's poorer farmers. Jack Kloppenburg, who wrote the classic book on the political economy of biotechnology in

---

[18] See Spielman, D.J., D.E. Kolady, A. Cavalieri, and N.C. Rao. 2014. 'The Seed and Agricultural Biotechnology Industries in India: An Analysis of Industry Structure, Competition, and Policy Options', *Food Policy* 45: 91–2.

[19] Venkateshvarlu, D. 2003. *Child Labour and Trans-national Seed Companies in Hybrid Cotton Seed Production in Andhra Pradesh*, Report commissioned by the Indian Committee on the Netherlands, available at http://www.indianet.nl/cotseed.html#contents; accessed on 24 July 2018.

[20] Peters, A.M. 2012. 'Bio-Informational Capitalism', *Thesis Eleven* 110(1): 105.

which he traced the evolution of plant breeders' rights in the USA and the shift away from public breeders to private breeders,[21] has highlighted, in a 2010 article, the potential for a bioinformatics that is based on open software.[22] What we are seeing in the context of biotechnology, just as we have seen in the context of dominant media content, is the availability of the 'copy', that is, products based on inclusive licensing arrangements that can be freely shared. Kloppenburg has been involved in the Open Source Seed Initiative (OSSI), whose principles have been derived from the open source software movement. This approach to seed is an attempt to make available both traditional and improved varieties of seed that are outside the control of proprietorial seed companies. It is also an attempt to re-foster traditions of seed exchange and propagation that have been marginalized by the relentless attempts to propertize seed. In the words of Kloppenburg:

> Like programmers, farmers have found their traditions of creativity and free exchange being challenged by the IPRs of the hegemonic 'permission culture' and have begun looking for ways not just to protect themselves from enclosure and dispossession, but also to reassert their own norms of reciprocity and distributed innovation. Moreover, farmers have potential allies in this endeavour who themselves are capable of bringing useful knowledge and significant material resources to bear. Although its capacity is being rapidly eroded, public plant science yet offers an institutional platform for developing the technical kernels needed to galvanize recruitment to the protected commons.[23]

Shemkus explains the operational basis for OSSI that is much like the terms described in Creative Commons licensing. Theirs is an intentional approach to 'free' the seed and make seed genes broadly available:

> To adapt the open source concept to seeds, OSSI decided to use a less formal pledge rather than a licensing system. Each packet of OSSI seeds sold will

---

[21] Kloppenburg, J.R. 2004. *First the Seed: The Political Economy of Plant Biotechnology*. Wisconsin: University of Wisconsin Press.

[22] Kloppenburg, J.R. 2010. 'Impeding Dispossession, Enabling Repossession: Biological Open Source and the Recovery of Seed Sovereignty', *Journal of Agrarian Change* 10(3): 367–88.

[23] Kloppenburg, 'Impeding Dispossession, Enabling Repossession', p. 377.

be printed with a statement that reads, in part, 'By opening this packet, you pledge that you will not restrict others' use of these seeds and their derivatives by patents, licenses, or any other means. You pledge that if you transfer these seeds or their derivatives you will acknowledge the source of these seeds and accompany your transfer with this pledge.'[24]

Transgenic Seed and Its Copy

Among the better-known examples of the seed as copy is the case of Navbharat Seeds Ltd, a company based in Ahmedabad, Gujarat, India. In 1998, it began selling Navbharat 151, a cotton seed that mysteriously killed the bollworm, a major cotton pest that was similarly targeted through Monsanto's seeds. In 2001, there was devastating bollworm infestation that wiped out all other varieties except Navbharat 151. Monsanto's Indian partner, Mahyco Corporation, became suspicious and tested the seed. They found that the seed was a Monsanto copy, although given that the Indian government did not allow seed patents at that time, they could not take this company to court or contest the company's defence that the seed's qualities were picked up by accident through cross-pollination from a Monsanto test plot.[25] Roy, Herring, and Geisler, in an article on the use of transgenic seeds in Gujarat, suggest that farmers traditionally are risk-averse and that, as such, they typically adopt a variety of seeds, including seeds that are of dubious provenance.[26] The fact that these are available and shared would suggest that it is extremely difficult for large agro-corporations such as Monsanto to impose their writ in countries like India, precisely because there are large gaps between regulatory regimes and local practices that are overladen by local politics.

[24] Shemkus, S. 2014. 'Fighting the Seed Monopoly: "We Want to Make Free Seed a Sort of Meme"', *The Guardian*, 3 May, available at http://www.theguardian.com/sustainable-business/seed-monopoly-free-seeds-farm-monsanto-dupont; accessed on 11 July 2018.

[25] See McGray, D. 2002. 'Biotech's Black Market', *Mother Jones*, September–October, available at http://www.motherjones.com/politics/2002/09/biotechs-black-market; accessed on 7 July 2018.

[26] Roy, D., R.J. Herring, and C.C. Geisler. 2007. 'Naturalising Transgenics: Official Seeds, Loose Seeds and Risk in the Decision Matric of Gujarati Cotton Farmers', *Journal of Development Studies* 43(1): 158–76.

Large and small farmers alike used saved and 'loose' seeds, officially approved and illegal transgenic seeds in 2002–03. The high price of official seeds distressed many; the uncertain quality of loose seeds worried others. Accordingly, they mixed and matched seeds according to their needs, not all of which were economic. Ongoing experimentation is an elemental part of their risk-management strategies; transgenics have simply added new possibilities to the mix. It is in this sense that most Gujarati cotton farmers have naturalized transgenics, fitting them into traditional strategies of conceptualizing and managing risk and assuring a livelihood.[27]

Herring, in perhaps the best researched article on 'stealth seeds' (defined as 'Stealth transgenics are saved, cross-bred, repackaged, sold, exchanged and planted in an anarchic agrarian capitalism that defies surveillance and control of firms and states'), has focused attention on the situation in India where transgenic cotton seeds are available, and also Brazil where equivalent 'soy' seed is available.[28] These have become widely distributed and have become difficult to control because of the nature of the digital 'copy'. As he points out:

> Farmers in Gujarat have embraced the agrarian anarcho-capitalism of stealth seeds. This outcome is not in the interest of seekers of innovator rents through state protection of intellectual property—in this case both Monsanto and Navbharat. The long development costs and time, mandated by a biosafety regime that is becoming global, are estimated to be about US$8 million for MMBL; these costs put official seeds at a price disadvantage. In Indian law, there is no restriction on farmer-to-farmer exchange of seeds. Farmer-generated transgenic hybrids that I have seen reference on the package the Indian Seed Act of 1966, Section 24, which protects this right explicitly.[29]

The case of copy transgenic seed reiterates the view expressed in this volume that digitally manipulated goods have many lives. It is difficult for companies to maintain a legal monopoly over seed in a context in which the majority of people depend on the availability of reliable and inexpensive

---

[27] Roy, Herring, and Geisler, 'Naturalising Transgenics', p. 171.

[28] Herring, R.J. 2007. 'Stealth Seeds: Bioproperty, Biosafety, Biopolitics', *Journal of Development Studies* 43(1): 130. See also Herring, R.J., 'Politics of Biotechnology: Ideas, Risk and Interest in Cases from India', *AgBioForum* 18(2): 145–55.

[29] Herring, 'Stealth Seeds', pp. 35–6.

seed. In the case of Navbharat Seeds Ltd, though it was proscribed by the government, its services contributed to the livelihoods of many thousands of small farmers in Gujarat. Once these farmers began to see the value of such seeds, it became impossible to restrict experimentation, exchange, and adaptation of seed, making it difficult for law enforcement to control this environment. An entire network of seed producers, companies, and agents serviced this need for illegal transgenics in Gujarat; and indeed, this has resulted in a parallel economy. Ramaswami, Pray, and Lalitha, in an article on this grey economy, highlight an issue that has been flagged in this volume, that is, the State is highly ambivalent in its response to the dictates of global capitalism and that it does in fact ignore, for reasons of political expediency, this parallel economy: 'As the production and distribution of illegal seeds is coordinated by a network of seed companies, seed producers, and seed dealers, enforcement is not difficult. The chain from seed plots to seed sales can be disrupted at any point. The lack of enforcement is an act of choice by the state governments.'[30] This reading is reinforced in a study by Murugkar, Ramaswami, and Shelar that Mahyco–Monsanto Biotech's dominance of the Bt cotton seed market has been 'challenged by illegal seeds, by alternative gene providers and by state policy.'[31]

The Turn towards Precision Agriculture

The politics of transgenic seed in India is one aspect of a gradual turn towards the expansion of the informational agricultural economy in India. While precision agriculture is yet to become a nationwide industry, it is clear that in the context of the agrarian crisis, the winners will be large farmers who have access to resources, land, and credit. This section will explore the informational basis for precision farming before dealing with the situation in India.

---

[30] Ramaswami, B., C.E. Pray, and N. Lalitha. 2011. 'The Spread of Illegal Transgenic Cotton Varieties in India: Biosafety Regulation, Monopoly, and Enforcement', *World Development* 40(1): 185.

[31] Murugkar, M., B. Ramaswami, and M. Shelar. 2007. 'Competition and Monopoly in the Indian Cotton Seed Market'. *Economic & Political Weekly* 42(37): 24, available at https://www.isid.ac.in/~bharat/Doc/ramaswami_epw2.pdf; accessed on 8 July 2018.

Monsanto, the global life sciences firm, bought The Climate Corporation for $930 million in 2013 precisely because it owned the code that could predict weather-related risk—key to farmers. It bought Precision Planters, a company that helps farmers make decision on fertilizers and crop densities, for $250 million in 2012. Both companies are part of Monsanto's Integrated Farming Services platform.[32] Monsanto's 2014 project, FieldScripts, 'uses seed science and precision agriculture equipment to accurately plant less seed in low potential areas and more seed in high potential areas. By recommending the best hybrids for the field and providing a prescription for variable rate planting, FieldScripts is able to help farmers maximize yield potential.'[33] This is called 'precision farming'. As Gabriel Lowy observes: 'Monsanto is at the crux of several major technology trends, including big data analytics, cloud and mobility: it brings together genetic and environment data from different databases, uses predictive analytics to create algorithms that are delivered on-demand via the cloud to mobile devices with data visualization tools on-board self-steering tractors.'[34]

Apart from the provision of inputs, companies such as DuPont have created digital farmer support systems such as Pioneer Field360, a subscription-based Web system that helps farmers plan their crop cycles based on field data with real-time agronomic and weather information. This is a case of IT helping with the development of 'precision agriculture': 'a management system that is location-specific, with data compiled from soils, crops, nutrients, pests, moisture, and/or yield, for optimum profitability, sustainability, and protection of the environment. Precision agriculture, also called "site-specific" or "prescription" farming, helps growers tailor their operations on a micro level, acre by acre, field by field, for maximum output and profit.'[35] There are other packages,

---

[32] Mitchell, D. 2013. 'Why Monsanto Spent $1 Billion on Weather Data', *Modern Farmer*, 2 October, available at http://modernfarmer.com/2013/10/monsanto-spent-1-billion-climate-data/.

[33] Monsanto. 'FieldScripts', available at http://www.monsanto.com/products/pages/fieldscripts.aspx; accessed on 11 July 2018.

[34] Lowy, G. 2013. 'Monsanto Is Bridging Genetics and Big Data Analytics', Tech-Tonics Advisors, 4 October, available at http://tech-tonicsadvisors.com/monsantos-analytics-field-dr; accessed on 22 July 2018.

[35] DuPont, 'The Science Behind Precision Agriculture', available at http://www.dupont.com/corporate-functions/our-approach/global-challenges/food/articles/the-science-behindprecisionagriculture.html; accessed on 11 July 2018.

such as Monsanto's FieldScripts and John Deere's FarmSight. Precision agriculture, based on an integration between spatial technologies and the management of cropping systems, includes possibilities for yield mapping and digital farm maps based on satellite imagery. Global Navigation Satellite Systems are now routinely used in crop, soil, and environmental testing and monitoring; terrain modelling; and the variable applications of inputs. Today, the tractor has become a data centre that is not only involved in the functionalities of furrowing and tilling but also in data analysis. So, it is not surprising that companies such as Monsanto are investing in a variety of technology companies, including data services and forecasting technologies. As is reported in a trade blog:

> With its growing analytics business, Monsanto is transforming itself into a data-driven enterprise while consolidating the market for precision farming technologies. Deeper insights into weather and planting can materially improve crop yields. This intelligence could then be correlated to biotechnology advances to further optimize production.[36]

IBM's Deep Thunder project, likewise, is a 24-hour weather-monitoring platform that can be linked to any firm's assets management strategy.[37]

Information technology, in other words, is playing a critical role in agribusiness and each section in the value chain has incorporated digital solutions. Agribusiness companies such as John Deere rely on bioinformatics tools that have become the basis for 'precision agriculture', and traders, food companies, and retailers use IT-tracking services and social media as part of their 'logistics' and engagement with customers. In other words, this is an example of the inter-sectorality and expansiveness of information technologies that are involved in remaking all productive sectors, including agriculture and the $4-trillion global food industry, in its image.

Precision farming clearly is an attempt to further commercialize global agriculture and incorporate farmers into the models of agricultural growth that have been popularized by global agribusinesses. It fits into the model of large capitalist farms, rather than small farms that are typically what farmers own throughout the world. In this sense, precision

[36] Lowy, G. 'Monsanto is Bridging Genetics and Big Data Analytics'.

[37] IBM, 'Deep Thunder', available at http://www-03.ibm.com/ibm/history/ibm100/us/en/icons/deepthunder/; accessed on 9 July 2018.

agriculture contributes to an accentuation of the divides between big and small farmers and to the growth in food volumes that is at the heart of the global food industry. Wolf and Buttel, in an article on the political economy of precision agriculture, conclude by stating that it basically preserves, extends, and deepens the hold of agribusinesses and their products—inclusive of seeds, pesticides, and fertilizers.[38] In their words:

> Precision farming is ... a high-technology means for rationalising the use of old—if not obsolete—farming inputs. It can be seen as a process of restructuring of agricultural practices so as to redress a set of problems caused by conventional IIA [industrial intensification of agriculture] practices, while simultaneously protecting and advancing the industrial structures, investments, and institutional arrangements premised on these practices.[39]

The global political economy of food security fits this model of agriculture given that greater productivity is seen to be the answer to the world's growing food insecurities. This script uses the reality of food insecurities in the developing world as the basis for technology-led solutions, from the Green Revolution to what some called the 'Evergreen Revolution' that is data driven. What is missing, according to the advocates of this new revolution in agriculture, is the lack of data (information)—an assumption that those of us involved in communication for social change have had to contest over many years. No mention is made of the need for people in the developing world to have enabling environments and access to land, fresh water, etc., which are required prior to data flows, making a difference to farmers. The challenge to feed the world will be powered by big data, enabled by trade with food mainly produced in the USA.[40] It is clear that 'B' versions of this model are being trialled in the developing world, such as in India, where Monsanto's Farm AgVisory Services uses a short message service (SMS) mobile-based platform to send daily

---

[38] Wolf, S.A. and F.H. Buttel. 1996. 'The Political Economy of Precision Farming', *American Journal of Agricultural Economics* 78(5): 1269–74.

[39] Wolf and Buttel, 'The Political Economy of Precision Farming', p. 1274.

[40] 'Special Report: Sustainability in the Age of Big Data', September 2014. Initiative for Global Environmental Leadership, Knowledge at Wharton, pp. 1–20, available at http://d1c25a6gwz7q5e.cloudfront.net/reports/2014-09-12-Sustainability-in-the-Age-of-Big-Data.pdf; accessed on 25 March 2019.

text messages to farmers on the entire cropping cycle, weather, etc.—in 7 languages and 16 states—particularly on corn, cotton, and vegetable farming.[41] However, proponents of food sovereignty contest this vision given that it completely sidesteps the issues of pricing, access, and more importantly, control over food production. In this sense, the crisis of capitalism is acutely witnessed in the current model of agricultural growth that is predicated on advances in biotechnology and genetic engineering based on ever-increasing controls over what is possible in global farming.

However, precision agriculture is not as benign as it sounds, as farmers who have incorporated such systems produce vast amounts of data, which in turn are fed back into the companies that manufacture such solutions, such as Monsanto and John Deere. Their use of private data on farms is a source of worry, given the issues with data privacy, security breaches, sale of data to third parties, and issues with data integration. There are also issues with IP given the lack of clarity over the ownership of the data that are generated. James Walter, in an article on the ownership of precision agriculture data, highlights some of these issues:

> Assuming the farmer maintains his relationship with the same supplier, he will be able to continue to build a database of information taken from his farm every year. The data gathered becomes more valuable as the number of total crop years increase because patterns develop and variables such as weather decrease in significance. The problem arises if the farmer ever decides to change suppliers, because if he does not own the database of information, he cannot take it with him. Effectively the supplier has him locked in, unless he has access to an electronic database and digital maps or can at least control its dissemination to another supplier.[42]

The operationalizing of Monsanto's FieldScripts is based on a close integration between the farmer who uses this software via an iPad application, the dealer who supplies it, and Monsanto that is involved in the

---

[41] 'Monsanto India to Expand Text Messaging Program'. 2014. Farm Chemicals International, available at http://www.farmchemicalsinternational. com/crop-inputs/monsanto-india-to-expand-sms-text-messaging-program/; accessed on 10 July 2018.

[42] Walter, J.R. 1997. 'A Brand New Harvest: Issues Regarding Precision Agriculture Data Ownership and Control', *Drake Journal of Agricultural Law* 2(2): 13.

analysis of farmer-generated information in a constant monitoring loop. In other words, surveillance as constant monitoring affects communities in different ways and, in the context of the USA, even farmers involved in large-scale farming do not trust the power of agribusinesses. In the USA, issues with vendor lock-in—again familiar to those of us who have been involved in writing on the political economy of information giants such as Microsoft—have led to the creation of initiatives such as the Open Ag Data Alliance (OADA) that is based on an open standards software platform. This initiative offers farmers possibilities for interoperability between their hardware and software and platform-neutral farm support. The OADA's objectives are as follows:

- will operate with a farmer-focused approach through a central guiding principle that *each farmer owns data generated or entered by the farmer*, their employees or by machines performing activities on their farm, will develop open reference implementations of data storage and transfer mechanisms with security and privacy protocols,
- will provide a forum for technical community discussion,
- will be led according to the processes of open standards projects that have built much of the Internet's networking, security, web and data standards with multiple university, individual and corporations participating (often while competing in the marketplace). Examples include the Internet Engineering Taskforce (IETF), World Wide Web Consortium (W3C) and The Apache Software Foundation, (which supports over 100 projects).
- will direct any financial contributions to a not-for-profit foundation whose purpose will be to enable open source projects in agriculture in service of the OADA mission (emphasis in original).[43]

Precision Farming/Agriculture in India

While the adoption of precision farming in India is still in its infancy, given India's investments in IT, there already are numerous initiatives run by both corporate establishments and various state governments that have invested in various types of precision farming. This area has been identified as a priority by the working groups of India: for example, the US Knowledge Initiative on Agriculture (KIA) that is part of a US $285

---

[43] Open Ag Data Alliance, 'Principles', available at http://openag.io/principles/; accessed on 11 July 2018.

million National Agricultural Innovation Project has been highlighted in the Government of Tamil Nadu's Precision Farming Project,[44] and is being explored by NGOs such as the M.S. Swaminathan Research Foundation based in Tamil Nadu and also by the private sector, including Tata Chemicals Ltd, part of the Tata Group of companies that is among the largest conglomerates in India. The KIA was signed in 2006 between President Bush and the then PM of India, Manmohan Singh, and the objective was to improve agricultural productivity, increase trade in such products, and contribute to what is called the 'Evergreen Revolution', the successor to Green Revolution in India.[45] The KIA, in the era of President Obama, was known simply as the 'Agricultural Dialogue'. Joint initiatives include a multimillion-dollar market information system, although it is also being used as a forum by the US government and key industries to pressure the Indian government to strengthen its regulatory environment and make it conducive to agricultural trade, biotechnology, and IP. The presence of Walmart, Monsanto, and Archer Daniels Midland on the board of KIA has fuelled speculations that its hidden objectives are to both open up the Indian agricultural sector to global businesses and harmonize its laws related to IP with global legislations.[46] Additionally, KIA has opened up the US agribusiness sector to a variety of central and state agricultural universities and research centres in India. The fact that KIA came into effect without proper consultation with farmers' groups was based on the participation of hand-picked central and state agricultural organizations on the Indian side and bypassed parliamentary approval highlights the nature of this agreement. Sridhar, in an article, underlines the impact of KIA on transgenic crops and the creation of regulatory regimes that are supportive of transgenic crops:

---

[44] Tamil Nadu Agricultural University. 'Tamil Nadu Precision Farming Project', available at http://agritech.tnau.ac.in/tnpfp-ENG/index.html; accessed on 22 July 2018.

[45] 'The US–India Agricultural Knowledge Initiative (AKI)', available at http://nifa.usda.gov/resource/us-india-agricultural-knowledge-initiative-aki; accessed on 2 July 2018.

[46] Parsai, G.2006. 'Wal-Mart and Monsanto on Indo-US Agricultural Initiative Board', *The Hindu*, 10 February 2006, available at http://www.thehindu.com/todays-paper/walmart-and-monsanto-on-indous-agriculture-initiative-board/article3176462.ece; accessed on 4 July 2018.

In September 2007, the Indian government announced that imports of transgenic food products no longer required approvals from the Genetic Engineering Approval Committee (GEAC), a body working under the Union Ministry of Environment and Forests. Although this decision was kept in abeyance in February 2008, the US had argued before the Committee on Technical Barriers to Trade (TBT) of the World Trade Organization (WTO) as early as 2006 that the very existence of the KIA reflected the central importance of biotechnology and, therefore, the need to have a non-discriminatory regime on transgenic varieties in India.[47]

The change in government in 2014 has led to a further strengthening of IP policy in India favourable to the US interests. In fact, a joint Indo-US IP Working Group was established in 2014 to discuss key issues, including IP in seed and the need for India to embrace the commodification of seed.[48]

The Indian anti-GM crusader Vandana Shiva, in a report, has highlighted Monsanto's role in KIA and beyond it:

India has signed a US–India Knowledge Initiative in Agriculture, with Monsanto on the Board. Individual states are also being pressured to sign agreements with Monsanto. One example is the Monsanto–Rajasthan Memorandum of Understanding, under which Monsanto would get intellectual property rights to all genetic resources, and to carry out research on indigenous seeds. It took a campaign by Navdanya and a 'Monsanto Quit India' *Bija Yatra* ('seed pilgrimage') to force the government of Rajasthan to cancel the MOU.[49]

A key consequence of corporate control over seed and this market has been its impact on peasant incomes, particularly those from Dalit (Untouchable) and *Adivasi* (tribal) backgrounds, and the further

[47] Sridhar, V. 2014. 'US Intervention in Indian Agriculture: The Case of the Knowledge Initiative in Agriculture', *Review of Agrarian Studies* 4(2): 97.

[48] Bhutani, S. 2016. 'Intellectual Property Rights Policy Fails to Address Farmers' Rights and Needs', *The Wire*, 30 May, available at https://thewire.in/39353/intellectual-property-rights-policy-fails-to-address-farmers-rights-and-needs/; accessed on 7 July 2018.

[49] Shiva, V. 2012. 'The Seed Emergency: The Threat to Food and Democracy', *Al Jazeera*, 12 February, available at http://www.aljazeera.com/indepth/opinion/2012/02/201224152439941847.html; accessed on 11 July 2018.

accentuation of uneven development. The Evergreen Revolution is based on the tighter integration of agro-chemical and biotechnology sectors with farming and while 'sustainable farming' has become part of the corporate mantra, it will, more likely than not, contribute to further degradation of the quality of soil that was one of the consequences of the Green Revolution.

Precision farming is seen as the answer to India's need to increase its food productivity from 241 million tonnes in 2010–11 to 480 million tonnes in 2050.[50] However, one of the major dilemmas in India is the shortage of land for farming and the fact that typically landholdings are small. Land fragmentation certainly does not help with the large-scale implementation of precision agriculture. However, there are many business opportunities for using and adapting precision agriculture to local needs and there are bound to be innovations in precision agriculture that are appropriate to landownership and economic status, such as the use of geographical information systems (GIS) and geographical positioning systems (GPS). However, the major adoptions of precision agriculture will be taken up by large farming establishments and in plantations. Mondal and Basu, in one of the more informative articles on precision agriculture in India, are confident that this can be applied to small holdings, although they also recommend that 'Two-hundred Agricultural Advanced Technology Parks (AATPs) should be developed in each region throughout the country, which will gather experience and develop methodology to apply PA precisely in region-wise format within the country.'[51] These AATPs will act as embryos for experimentation with precision agriculture—a thinking that is very closely aligned to the 'diffusion of innovations' tradition associated with Everett Rogers and other US academics. How land will be acquired to establish AATPs has not been dealt with in this article, although it would seem that this is a critical issue in the context of present-day India.

[50] Madhavan, N. 2012. 'Precision Farming Can Solve India's Many Farm Ills and Address Many of Its Food Security Issues', *Business Today*, 19 August, available at https://www.businesstoday.in/magazine/special/innovation-agri-culture-precision-farming/story/186624.html.

[51] Mondal, P. and M. Basu. 2009. 'Adoption of Precision Agriculture Technologies in India and in Some Developing Countries: Scope, Present Status and Strategies', *Progress in Natural Science* 19(6): 664.

Howlett and Velkar's study of the diffusion of ideas within the Tamil Nadu Precision Farming Project that involved 400 farmers indicates that its 'success' was due to the availability of subsidies to cover the purchase of 'fertigation equipment' and 'the cost of installation'.[52] Following the withdrawal of subsidies, it was unclear as to whether the majority of farmers would continue investing in high-technology interventions. The authors, however, note in the following interviews with some of the farmers that: 'The largest extension mentioned was by G., who said that he and some other farmers had bought equipment that would allow them to extend precision farming to another twenty acres.'[53] This confirms the fact that precision agriculture works best when farmers have access to more than average landholdings. Precision agriculture really is about technology that enables the prediction of external variables, such as the weather, and the application of precise inputs, such as GIS-based pesticide applications.[54] It is hoped that IT-based applications will improve efficiency and predictability and result in greater productivity. The Evergreen Revolution, it would seem, is the next stage of the Green Revolution, although very few of the studies on precision agriculture deal with the consequences of the ever-greater industrialization of agriculture for both people and the environment.

Agricultural MNCs in India are involved in a wide range of initiatives and investments in both production and policy. While the Indian agricultural environment certainly poses challenges to MNCs, they have, via their control over hybrid seed, steadily strengthened their hold over Indian farmers. Since marketed hybrid seeds are the 'property' of MNCs, the more farmers use industrialized hybrids, the more opportunities there are to control the farmer's use of seeds. In 2008–9, private seed companies accounted for 64 per cent of the 460 applications for Protection of Plant Varieties-certified novel hybrid seeds. While hybrid cotton accounted for 65 per cent of the seed being used, maize stood at 8 per cent, and

---

[52] Howlett, P. and A. Velkar. 2010. 'Technology Transfer and Travelling Facts: A Perspective from Indian Agriculture', in P. Howlett and M.S. Morgan (eds), *How Well Do Facts Travel?: The Dissemination of Reliable Knowledge*. New York: Cambridge University Press, p. 279.

[53] Howlett and Velkar, 'Technology Transfer and Travelling Facts', p. 293.

[54] Tiwari, A. and P.K. Jaga. 2012. 'Precision Farming in India: A Review', *Outlook on Agriculture* 41(2): 139–43.

rice at 6 per cent.[55] In other words, the untapped market for hybrid maize and rice remains massive, although the more the seed companies increase their market share, the less the farmers will be able to maintain their independence related to the use of seed. As Spielman et al. have observed: 'the relatively recent development of a private seed industry in India has meant that for many staple crops, particularly rice and wheat, farmers are still making the transition from saved seed, seed exchanges with neighbours, or purchases from public seed suppliers to buying seed from private companies.'[56] Irrespective of the prevalence of the copy in transgenic seed, it is this 'tradition' of sharing and propagating seed that is at stake. The steady pressure to deregulate the pricing of seed and its subsidization by the State has been underscored by the US–India Business Council, who have identified the capping of prices for seed by the State as the major issue: 'non market-based pricing as one of the most significant disincentives to the commercialization of new biotech seeds by global seed firms in India.'[57]

While 'stealth' seeds have arguably democratized access to affordable transgenic seed, it is also another instance of the digital as 'copy' and the difficulties of curbing its propagation, circulation, and use in countries that are as large and diverse as India. However, despite its potential to close the access gap to transgenic seed, there is as yet little information available on whether or not diluted transgenic seed have any impact on the environment and health. If stealth seeds are a form of cultural piracy,

[55] Kolady, D., D.J. Spielman, and A.J. Cavalieri. 2010. 'Intellectual Property Rights, Private Investments in Research and Productity Growth in Indian Agriculture: A Review of Evidence and Options, International Food Policy Research Institute, IFPRI Discussion Paper, 01031. November, pp. 1–48, available at https://core.ac.uk/download/pdf/6237679.pdf; accessed on 25 March 2019.

[56] Spielman, D.J., D. Kolady, A. Cavaleiri, and N.C. Rao. 2011.'The Seed and Agricultural Biotechnology Industry in India: An Analysis of Industry Structure, Competition and Policy Options', International Food Policy Research Institute Discussion Paper 01103, July, p. 91.

[57] Linton, K. and M. Torsekar. 2009. 'Innovation in Biotechnology in Seed: Public and Private Initiatives in India and China', 5 October, available at http://www.brookings.edu/~/media/events/2009/10/23%20china/ex2_paper3_linton.pdf; accessed on 5 July 2018.

there is also a fairly large anti-GM movement in India whose agenda includes the need for seed sovereignty. This movement includes a diverse array of actors from across India who are committed to preserving the seed sovereignty of India's farmers. These include: the Beej Bachao Andolan (Save the Seed Movement in Uttarakhand, north India); women farmers' networks involved in the preservation and propagation of indigenous seed, such as at the Deccan Development Society (AP), the Kokopelli Seed Foundation (Puducherry), and Vandana Shiva's Navdanya (Nine Seeds) movement; and the Karnataka Rajya Raitha Sangha (KRRS), a farmer's movement that began a crusade against Monsanto, the Gene Campaign, among many other initiatives. These movements are, very interestingly, also backed by a range of political groups representing the left, the centre, and the right in India. While the left in India has traditionally supported the rights of farmers, right-wing groups such as the RSS, which provide the ideological basis for Hindu nationalism, also count small farmers among its members. The RSS and at least some of its allies are ideologically committed to strengthening India's sovereign status and are distrustful of MNCs and India's embrace of globalized futures. This stance is opposite to the pro-business and pro-MNC orientation of the current ruling party in India, the BJP, which is the political wing of Hindu nationalism in India. Roy, in an article on the anti-GM movement in India, recalls one such conflict within the Sangh Parivar:

> In June 2014, a new government was formed by the right-wing *Bharatiya Janata Party* (BJP) in New Delhi. There was some fear among anti-GM activists that the new government would continue with the previous government's stance on field trials of GM crops. However, due to pressure from the right-wing organizations such as the *Swadeshi Jagaran Manch* (SJM) and the *Bharatiya Kisan Sangh* (BKS), the new government placed a hold in July 2014 on the field trials of fifteen GM crops.[58]

Monsanto's announcement of its plans to curtail the introduction of the successor to the modified cotton seed Bollgard II in 2016 was, to a large extent, a response to what the company saw as a determined effort by the Indian state to slash royalties to sellers of GM seeds, in particular

---

[58] Roy, D. 2015. 'Contesting Corporate Transgenic Crops in a Semi-peripheral Context: The Case of the Anti-GM Movement in India', *Journal of World Systems Research* 21(1): 98.

Monsanto that controlled 70 per cent of the market share, and introduce compulsory licensing of modified seed. Behind the Indian state's response was the hand of the RSS and its farmer's union, the BKS, along with a long-standing RSS member, the agriculture minister, Radha Mohan Singh.[59]

This 'double movement' is characterized, on the one hand, by the advent of neo-liberal agricultural policies and support for IP, in particular plant breeders' rights in seed and precision farming, and on the other, the growth of farmers' movements all over India that are committed to preserving farmers' rights to seed and food sovereignty. While the pro-business line taken by the BJP was key to its electoral success, its survival will be dependent on its adeptness in supporting both populist and pro-business agendas on agriculture. Given its pro-business credentials and record, it would seem that despite the existence of strong anti-GM movements in India, it will not be able to resist the overtures and pressures from the US lobbies to further modernize the agricultural sector in India.

[59] 'How an Indian Cotton Seed Producer Took on US Giant Monsanto'. 2017, *Business Today*, 28 March, available at https://www.businesstoday.in/current/economy-politics/how-an-indian-cotton-producer-took-on-seed-giant-monsanto/story/248852.html; accessed on 11 July 2018.

5

The Politics and Geopolitics of Internet Governance

The issue of Internet governance (IG) became a globally mediated discourse in the aftermath of the Snowden revelations of mass spying by the NSA and the complicity of key IT, software and hardware companies, and search engines in the surveillance of its publics. In the face of strong criticisms from leaders in both the developed and developing world, the 'reform' of ICANN along with the US government's voluntary withdrawal of its stewardship of Internet Assigned Numbers Authority (IANA) led to multiple conversations related to the democratization of IG. These conversations took place at United Nations (UN) agencies and in technical forums, government bodies, and civil society. The Just Net Coalition, for example, was founded in 2014 for providing a forum for consistent civil society discussions on various aspects of the democratization of IG. Following the announcement by the National Telecommunications and Information Administration (NTIA) on 12 March 2014 regarding the transition of IANA functions to the global Internet community, a number of important meetings were held, including NETmundial (23–4 March) organized by the Brazilian government, the International Telecommunication Union's (ITU) World Telecommunication Development Conference (30 March–10 April), the ITU Plenipotentiary Conference (20 October–7 November), World Summit on the Information Society (WSIS+10, 15–16 December), along with a number of meetings organized by the ICANN. While the apparent willingness of the US government to relinquish its hold over IG earned it media space, this was a controversial decision, and it would seem from subsequent conversations that the US government is keen on giving IG a new face but not on loosening its hold over core aspects of IG, including de facto control over the majority of its root servers.

This chapter explores the relationship between the Indian State and IG at both global and local levels. It takes the position that far from being a marginal player in IG, the State in India does play an important role in the shaping of IG at a local level, although this commitment is less apparent at an international level. The chapter focuses on three aspects. First, it explores the history of ICANN and its continuing relationship with the US government. Second, it explores the rather ambivalent stance taken by the Indian state towards global IG. The State has exhibited, at various times, its support for multilateral solutions to IG as opposed to the multistakeholder status quo. I argue that this ambivalence is a reflection of the State wanting maximum sovereignty related to the governance of the Internet, although it also is a reflection of turf wars between the Ministry of External Affairs (MEA), which takes a statist foreign policy line, and the DeitY, which is more open to the participation of the business sector and, to some extent, civil society. Third, this chapter explores the internal experience of IG, in particular the Indian state's multifaceted role in overseeing IG at local levels and the ways in which multistakeholderism (MSH) is playing out. At a domestic level, the State is keenly involved in IG issues such as data retention and encryption, and has even produced a position paper on the 'Internet of things'. The chapter argues that the State's veering towards MSH in the matter of IG is a reflection of its close relationship with the US and its neo-liberal commitments to growth with security.

The Internet and State Sovereignty

As the Internet has become an indispensable tool in business, politics, trade, networking, and communication, its governance has become an equally important concern. Internet governance, pre-ICANN, was an 'informal' arrangement characterized by a relatively open approach to the creation of standards and rules that reflected the libertarian beginnings of the Internet. Jan Postel's stewardship of 'assigned names and numbers' in the 1980s and 1990s exemplified an attitude that was based on trust in an individual to run the Internet. That was, however, in the days when the Internet was still in its infancy, when the Transmission Control Protocol/Internet Protocol (TCP/IP) that facilitates interconnectivity between devices did not need to deal with an exponential increase in the connectivity potential of devices and technologies as, for instance, third-generation (3G) and fourth-generation (4G) mobile phones, Wi-Fi, cloud computing, the Internet of things, and mesh technologies, to name just a few of the device and technology advances in recent years. The extraordinary potential of the Internet as a platform for commercial trade, State surveillance, and social networking has added to the urgency of deliberations on the governance of the Internet. The fact that citizens are simultaneously netizens as well remains a hugely problematic issue for governments, given that the Internet has both stymied and muddied the jurisdictional remit of the State over its citizens. The control of a netizen is, as governments have learned, difficult because individuals now have the capacity to make a distributed impact within a networked, dispersed system that is just as easily transnational as it is local. The current furore related to the net-based recruitment patterns of terrorist organizations, debates related to net neutrality, along with attempts to outlaw peer-to-peer file-sharing cultures that began with Napster and continued with Kazaa and Nutella, exemplify some of the protractive issues that continue to dog the Internet and highlight the growing importance of its governance for the State, business, and civil society.

In an article written in 1996, Johnson and Post describe the State's control over physical space such as territory as the remit of national law.[1] Such control gives the State power over people who are located within their borders, an ability to control the 'effects' of individual and group behaviour within that territory, the assent and legitimacy to rule over the governed in that territory, and gives 'notice' that another set of rules determine the behaviour of people when they cross the border into the territory of another sovereign state. However, cyberspace is profoundly disruptive to a State's identity as lawmaker, and law enforcer, because the terrain of the cyberspace is the transnational:

> The rise of the global computer network is destroying the link between geographical location and: (1) the power of local governments to assert control over online behaviour; (2) the effects of online behaviour on individuals or

[1] Johnson, D.R. and D. Post. 1996. 'Law and Borders: The Rise of Law in Cyberspace', *First Monday* 1(1): 1025, available at http://firstmonday.org/ojs/index.php/fm/article/view/468/389; accessed on 22 March 2018.

things; (3) the legitimacy of the efforts of a local sovereign to enforce rules applicable to global phenomena; and (4) the ability of physical location to give notice of which sets of rules apply. The Net thus radically subverts a system of rule-making based on borders between physical spaces, at least with respect to the claim that cyberspace should naturally be governed by territorially defined rule.[2]

Arguably, for a traditionally top-heavy state such as India, the challenges posed by the Internet are complex, and it is only of late that it has begun to realize the need for public policy related to the Internet. The State, in spite of its democratic credentials and its support for constitutional rights, has historically had an ambivalent relationship with its citizens, especially in the matter of their information and communication rights. While the post-liberal mediascape and the continuing regulatory vacuum does suggest a State that has withdrawn from its control of communication, its continuing control over community broadcasting, attempts to control the cable and satellite sector through support for a handful of large multi-service providers as opposed to the 60,000 independent cable operators, extension of the UID scheme, and censorship of the Internet (as mentioned earlier, the Indian government issues more takedown orders to Facebook and Google than any other country) do indicate that the Indian state is not ready to opt for a soft-touch approach to the control over communication and information. I have elsewhere used the term 'ambivalent state' to describe the Indian state given that it is capable of extraordinary social inclusion via welfare projects, such as the National Rural Employment Guarantee Act, 2005, that guarantees 100 days of employment in public works for every Indian citizen below a certain economic threshold and public-sector software that is based on the principle of FOSS, along with extraordinary exclusion through turning a blind eye to the reality of extreme poverty and denying 'suspect' populations in regions such as the north-east India their right to communication. This sentiment is reflected in Akhil Gupta's observation that: 'Popular sovereignty' in India 'takes the form of inclusion and unspeakable violence; forms of belonging coexist with the production of bare life'.[3]

[2] Johnson and Post, 'Law and Borders', p. 1025.
[3] Gupta, A. 2012. *Red Tape: Bureaucracy, Structural Violence and Poverty in India.* Hyderabad: Orient Blackswan, p. 18.

Issues related to IG need to be viewed against the background of the steady spread of Internet connectivity and yearly increases in the mobile phone subscriber base that was in the region of 933 million in 2014,[4] up from 700 million in 2011. The projected connectivity at the end of 2014 was 243 million Indians, the majority through mobile phones,[5] although this figure does not account for connectivity through the 'jugaad' (informal, disruptive) cultures that are serviced by mobile-phone kiosks throughout the length and breadth of India. Rangaswamy and Densmore describe this economy and the deals that help provide Internet access to poorer communities:

> Micro pre-pay and deal-a-day Internet plans are dexterously used to juggle with resource constraints and the need to persist with the internet. Sharing internet time among friends to searching appropriate software, optimizing them for the mobile phone, and tips and tricks circulating among resource crunched users to expand talk time and internet downloads are a few bottom-up examples of extending and expanding use. Mobile phone and PC-based repair ecologies in the slums of urban India, employing informal and even illegal channels, are actively expanding services to maintain and repair a variety of ICTs for the resource poor populations.[6]

Other initiatives are adding momentum to and hastening Internet connectivity in India, such as Facebook and Reliance Communication's launch of Internet.Org, which allows access to 33 websites (an Internet starter) that can be accessed from any Reliance mobile or tablet for free, although it has also been critiqued for ignoring the principle of net

[4] 'India's Telephone Subscriber Base Rises to 933 Million'. 2014. *The Times of India*, 12 May, available at http://timesofindia.indiatimes.com/tech/tech-news/Indias-telecom-subscriber-base-rises-to-933-million/articleshow/35024488.cms; accessed on 21 March 2018.

[5] 'With 243 Million Users by 2014, India to Beat US by Reach: Study'. 2013. *The Times of India*, 14 November, available at http://timesofindia.indiatimes.com/tech/tech-news/With-243-million-users-by-2014-India-to-beat-US-in-internet-reach-Study/articleshow/25719512.cms; accessed on 22 March 2018.

[6] Rangaswamy, N. and M. Densmore. 2013. 'Understanding Jugaad: ICTD and the Tensions of Appropriation, Innovation and Utility', Paper presented at the ICTD 2013, Cape Town, South Africa, 7–10 December, p. 4, available at http://research.microsoft.com/pubs/205396/ICTD2013-Rangaswamy-Jugaad.pdf; accessed on 11 March 2018.

neutrality. This increase in Internet users is a major boost to search engines, social networking firms, and e-business and commercial connectivity operators, but it also presents the government with major issues related to the control of population on the net.

The ICANN, the USA, and the Politics of MSH

Governance is often used to describe multilateral or multistakeholder processes based on negotiations aimed at the creation of policy pathways on a range of issues, from the nature of trade in agriculture to cultural policy at national or international levels. It is, as James Rosenau has observed:

> not synonymous with government. Both refer to purposive behaviour, to goal-oriented activities, to systems of rule; but government suggests activities that are backed by formal authority, by police powers to insure the implementation of duly constituted policies, whereas governance refers to activities backed by shared goals that may or may not derive from legal and formally prescribed responsibilities and that do not necessarily rely on police powers to overcome defiance and attain compliance.[7]

The expansive nature of governance that is not backed by legal statutes or formal authority can result in issues linked to transparency and accountability.[8] In the case of ICANN, we have an instance of a non-profit organization established in 1998 as a public benefit corporation under the California Non-Profit Public Benefit Corporation Law, at the behest of IANA that had, until 1998, the responsibility to allocate domain names and IP addresses. The ICANN is answerable to the US Department

---

[7] Rosenau, J.N. 1992. 'Governance, Order and Change in World Politics', in James N. Rosenau and Ernst-Otto Czempiel (eds), *Governance without Government: Order and Change in World Politics*, Cambridge: Cambridge University Press, p. 4.

[8] See Hunter, D. 2003. 'ICANN and the Concept of Democratic Deficit', *Loyola of Los Angeles Law Review* 36(3/4): 1149–84; Marlin-Bennett, R. 2001. 'ICANN and Democracy: Contradictions and Possibilities', *The Journal of Policy, Regulation and Strategy for Telecommunications, Information and Media* 3(4): 299–311; and Weber, R.H. 2008. 'Transparency and the Governance of the Internet', *Computer Law & Security Report* 24(4): 342–8.

of Commerce with which it has an MoU that has been amended on six occasions. As Jeremy Malcolm points out in his voluminous account of multistakeholder governance and the Internet Governance Forum (IGF), the inception of ICANN was marred by controversy.[9] The Internet International Ad Hoc Committee (IHAC), which included the ITU, WIPO, and other organizations, had pushed for competition in the allocation of generic top-level domain (gTLD) names and a separation between the function of the registrar of these names and the registry. They created a gTLD MoU that was criticized by those who were not keen on UN agencies having a role in IG. Another group, the International Forum on the White Paper (IFWP), based largely on IHAC members, continued exploring the basis for a new body that would take on the governance of the Internet. However, it was IANA's proposal for ICANN that eventually succeeded and Malcom has observed that 'IANA's high-handed circumvention of the IFWP process caused significant dissent.'[10]

The US government's approach to the Internet has been founded on the primacy of the market and the flow of online commerce; and this continues to be their stance in the post-Snowden era. Singh cites five principles that sum up the approach of the USA to IG: '1) private sector leadership ... 2) minimal government intervention, 3) minimalist, consistent and simple legal environment, 4) recognition of the decentralised nature and tradition of bottom-up governance of the Internet, and 5) consistent global governance principles to create predictable environments regardless of jurisdictions.'[11]

The US government has also taken an unequivocal stance on the location of ICANN with the latest IANA contract, signed on 1 October 2012, stipulating that ICANN be 'a wholly US-owned firm operating, incorporated and organised under US law.'[12] The IANA contract remains

[9] Malcolm, J. 2008. *Multi-stakeholder Governance and the Internet Governance Forum*. Perth: Terminus Press.

[10] Malcolm, *Multi-stakeholder Governance and the Internet Governance Forum*, p. 37.

[11] Singh, J.P. 2009. 'Multilateral Approaches to Deliberating Internet Governance', *Policy & Internet* 1(1): 100.

[12] Weitzenboeck, E.M. 2014. 'Hybrid Net: The Regulatory Framework of ICANN and the DNS', *International Journal of Law and Information Technology* 22(1): 53.

a key tool for the oversight of ICANN by the US government.[13] This contract is key to ICANN since it gives it the sole right to deal with top-level domain (TLD) names, such as .com. In fact, companies based outside of the USA were not allowed to bid for the IANA contract.

The ICANN is entrusted with the management of three specific functions: (i) allocating domain names, in particular the gTLD and country code top-level domain (ccTLD) name system management; (ii) establishing Internet Protocols; and (iii) some of the functions of root server management. It has traditionally awarded contracts for top domain names to US-based organizations, such as Verisign that handles the .com domain name and the Public Interest Registry that handles the .org domain name. Both organizations are based in Virginia and Verisign also manages the central root server that is based in Dulles, Virginia. However, the governance of the Internet includes involvement in the creation of rules, standards of behaviour (norms), maintenance, and expansion of its architecture, networks, the information economy, and public policy. While ICANN is known for its involvement in domain name and its Internet Protocol-related functions, there are a host of organizations involved in the governance of the technologies of the Internet. These include the Internet Architecture Board (IAB), the Internet Engineering Task Force (IETF), the IGF, the International Organization for Standardization Maintenance Agency, the World Wide Web Consortium (W3C), and other bodies. It is unclear whether these bodies, such as the IGF that was established immediately after the Tunis phase of the WSIS in 2005, have contributed in any substantive manner to the shaping of Internet policy, or whether they have remained a contained platform for an MSH that is just not replicable in any of the Internet's governance bodies, including ICANN. In other words, it is a moot point as to whether or not deliberations at the IGF have impacted on mainstream IG. Van Eeten and Mueller suggest that the IGF is a site for the celebration of IG as a discourse:

> The UN IGF ... is assumed to be relevant to the governance of the Internet because that is where people come together to talk about Internet governance.

[13] See McGillivray, K. 2014. 'Give It Away Now? Renewal of the IANA Functions Contract and Its Role in Internet Governance', *International Journal of Law and Information Technology* 22(1): 3–26.

This obscures the rather painful fact that most of the actors with operational control over the Internet resources are absent from it ... The IGF has produced no collective resolutions, let alone binding agreements or decisions, and even if it did, these would have no commitment power over the actors actually operating the Internet.[14]

In other words, at the level of public policy, the IGF has not made any substantive contributions except facilitating the knowledge base on the governance of the Internet among stakeholders. Among the most salient issues following the aftermath of the Snowden revelations was ICANN's role in global IG and that of the US government relinquishing its oversight of IANA. Despite all the rhetoric, it would seem the case that the US influence will continue to play a critical role in the governance of the Internet.

For the most part, global policy on a host of issues is carried out through multilateral means—meaning conversations predominantly between national governments, although the private sector has increasingly been involved in some of these conversations. Multilateral talks are the preferred means at a UN level, at the WTO, and at other international forums. The ICANN, however, is different in the sense that, from the very beginning, the accent has been on multistakeholder processes, reflecting perhaps the origins of the Internet that were shaped by different actors, from the US military to the private sector and civil society. In the case of ICANN, there have been major issues regarding the gaps between its principles and practices of MSH. One could argue that this was a doomed process from the very beginning given that the US government had exclusive oversight over ICANN. The establishment of the IGF was to some extent a counter to this exclusivity and it provided the space for wide ranging conversations between the world's internet stakeholders. The ICANN website offers an explanation for the variety of MSH followed:

ICANN's inclusive approach treats the public sector, the private sector, and technical experts as peers. In the ICANN community, you'll find registries, registrars, Internet Service Providers (ISPs), intellectual property advocates, commercial and business interests, non-commercial and non-profit interests, representation from more than 100 governments, and a global array of

[14] Van Eeten, M.J.G. and M. Mueller. 2012. 'Where Is the Governance in Internet Governance?', *New Media & Society* 15(5): 728.

individual Internet users. All points of view receive consideration on their own merits. ICANN's fundamental belief is that all users of the Internet deserve a say in how it is run.[15]

This laudable goal is translated into a number of ICANN-related committees that provide a forum for stakeholder discussions. A good example of such a group is ICANN's At-Large Advisory Committee (ALAC) that consists of 180 members representing ordinary Internet users from around 80 countries. At their meeting held in London in June 2014, one of the themes discussed was MSH. It is fascinating to note that one of the objectives of this exercise was to validate this form of MSH against what they perceived to be the regressive mood at NETmundial (a meeting hosted by the Brazilian government in 2014, in the aftermath of Snowden revelations, at which the issue of alternative to IG was discussed) and attempts by governments to establish a statist model for IG. As the draft of this group's report states:

> Of particular interest and concern to the Group is the role of governments within the context of the MSMs (multi stakeholder models). A number of government statements during the NetMundial meeting indicated a discomfort with MSMs and a desire to revert to intra-governmental policy making, either through the ITU or a strict interpretation of WSIS declarations. Many members of the Group had encountered situations in which governments asserted that they believe that they are above the MSMs. The argument that is usually highlighted here is the fact that democratically elected bodies claim to represent the public interest. However, the Group felt not all governments are democratically elected, nor the fact that all act on public interest.[16]

The need to be 'vigilant' to any takeover of IG is a dominant thread in liberal arguments supportive of the IG status quo.

While there definitely are many governments from around the world (such as the BRICS nations) that are keen to explore other platforms for

---

[15] ICANN, 'Resources', available at https://www.icann.org/resources/pages/welcome-2012-02-25-en; accessed on 11 March 2018.

[16] 'ATLAS II: Thematic Group of the Future of Multistakeholder Models', available at https://community.icann.org/display/atlarge/ATLAS+II+Thematic+Group+on+the+Future+of+Multistakeholder+Models; accessed on 26 March 2019.

IG that are different from ICANN and some that are keen for the ITU or some other UN-related body to take over IG, the valorization of ICANN's version of MSH is interesting precisely because it strongly suggests that governments should not be given primary responsibility for the creation of public policy related to the Internet and that this should remain a prerogative of a multistakeholder forum. The ICANN's concession to democracy has been to create a number of lower-level advisory groups, such as the Government Advisory Committee (GAC) whose brief is limited to providing advice and ALAC that is involved in a limited type of what one might call 'deliberative democracy'. While the GAC's power to shape IG policy is arguable, the perception that it is toothless has certainly informed the Indian government's MEA preference for a multilateral model, exemplified by its support for an alternative body—the Committee for Internet Related Policies (CIRP)—at a UN meeting in 2011.[17] While the involvement of civil society in IG is laudable, there is little evidence either at ALAC, ICANN, or IGF that the civil society groups involved have been able to contribute to policymaking. In both cases, one can argue that these groupings are working in splendid isolation from the mainstream of IG. This is arguably a form of corporatism: 'a system that gives a variety of functional interest groups ... direct representation in the political system, defusing conflict among them and creating instead broad consensus on policies.'[18] Ottaway, however, is sanguine about the impact of such processes: 'despite the claims that tripartite agreements will introduce greater democracy in the realm of global governance, it is doubtful that close cooperation between essentially unrepresentative organisations—international organisations, unaccountable NGOs and large transnational corporations—will do much to ensure better protection for, and better representation of, the interests of populations affected by global policies.'[19]

[17] Julka, H. 2012. 'Internet Censorship: India to Push for Internet Regulation at the United Nations', *The Times of India*, 24 August, available at http://articles.economictimes.indiatimes.com/2012-08-24/news/33366607_1_internet-governance-internet-censorship-objectionable-content; accessed on 11 March 2018.

[18] Ottaway, M. 2001. 'Corporatism Goes Global: International Organisations, Nongovernmental Organisation Networks and Transnational Business', *Global Governance* 7(3): 268.

[19] Ottoway, 'Corporatism Goes Global', p. 266.

One can argue that in countries such as India in which large numbers of the population are completely dependent on government welfare, the access to the Internet for majority of its population can only be provided through the State's public policy commitments to information for all. Neither the market nor civil society have the commitment or reach to attain such objectives. This position is favoured by Parminder Jeet Singh, a prominent Internet activist from the Bengaluru-based NGO, IT for Change, who declared in a personal interview that he was against:

> a certain type of internet exceptionalism. It is being argued that the internet is some kind of uniquely different place in which the typical ways of public policy making do not work—and I don't agree with it. When you talk about public policy making for example in health, education—these cannot be developed by corporations sitting on the table at the same level as governments if that is what is meant by MS. If MS is about deep consultative democracy then we are for it—but MS is a deliberate smoke screen used to defend the status quo … those who make public policy can only be made by those who represent the public—perfect or imperfectly.[20]

In other words, if the Internet is seen as a public resource, then it is difficult for the State to be denied substantive involvements in IG given that neither the private sector nor civil society has the resources, will, or mandate to extend connectivity to all. A Declaration of the Committee of Ministers on ICANN, Human Rights and the Rule of Law, by the Council of Europe, includes an implicit role in its governance by the State:

> The Internet is a global resource which has public service value and should be managed in the public interest. People, communities, public authorities and private entities rely on the Internet for their activities and have a legitimate expectation that the Internet will remain one unfragmented network and that its services will be accessible, provided without discrimination, affordable, secure, reliable and continuous.[21]

[20] Personal interview with Parminder Jeet Singh, IT for Change, on 2 April 2015.

[21] Council of Europe. 2015. 'Declaration of the Committee of Ministers on ICANN, Human Rights and the Rule of Law', 3 June, available at https://wcd.coe.int/ViewDoc.jsp?id=2328763&Site=CM; accessed on 11 June 2018.

I think that the concept of MSH favoured by the US needs to be reassessed precisely because it is based on the principle of 'equal say' for all stakeholders. It is clear that at the core of the notion of MSH is the belief in the value of 'free' and unregulated information flows that was a mainstay of the US during the heady days and debates related to the New World Information and Communication Order (NWICO) in the late 1970s and early 1980s.[22] Multistakeholderism, with its promise of participation, voice, and dialogue as the basis for a consensus-based approach to policymaking, is alluring precisely because it suggests participatory democracy in action. However, while it sounds good in theory, it is only reasonable to expect that the major Internet companies, because of their power and influence, will play a more important role in the creation of policy than civil society, which in any case consists of organizations who do not have a singular vision on the public policy aspect of IG. It is to be expected that these major players will focus their efforts on establishing the Internet as a platform that will generate maximum economic returns supported by preferential tax regimes. Michael Gurstein, in an article, articulates one of the better critiques of the MSH currently being followed in the context of IG:

> In a multistakeholder governance regime ... the Internet giants will presumably be equal partners/stakeholders in the determination of matters of Internet regulation, taxation and the possible (re)allocation of overall benefits, i.e. those matters which are of direct financial concern to themselves and their shareholders. And these determinations will be taking place in policy contexts where there are no obvious champions/stakeholders representing the broad global public interest. That such an arrangement is directly supportive of US and other developed-country interests and the interests of dominant Internet corporations, i.e., those most actively lobbying for the multi-stakeholder model, is clearly no accident.[23]

This critique of the form that MSH has taken in ICANN does not imply that the multilateral model is better. What I have tried to highlight here is the fact that there is a definite bias towards a specific

[22] See Schiller, H.I. 1974. 'Freedom from the "Free Flow"', *Journal of Communication* 24(1): 110–17.

[23] Gurstein, M. 2014.'The Multistakeholder Model and Neoliberalism', *Third World Resurgence* 287/288: 22–24.

multistakeholder approach in the matter of IG, although when one takes a hard look at who is involved in the governance of the Internet, it includes organizations like the WIPO and the WTO, who are multilateral in their approach, along with an assorted group of regional bodies, such as the EU, the US Department of Commerce, and a number of private enterprises. The WTO, for example, has settled two trade disputes related to the Internet: Internet gambling (2007) and China's state trading rights on audiovisual products and services.[24] It is also involved in issues around digital copyright and is actively involved in the export of US IP standards to the rest of the world. This heterogeneity in terms of IG can, on the one hand, be seen as the best solution for a platform that requires a range of approaches related to its governance. However, on the other hand, it does provide opportunities for countries like the USA to take maximum advantage of the different approaches to IG. Richard Hill has observed that: 'what is really being argued is that certain particular Internet governance matters should be decided by multistakeholder bodies.' These matters, of course, relate to critical issues related to IG: 'Perhaps not coincidentally, the matters in question are those that are handled by the existing prominent Internet bodies such as ICANN, IETF, W3C, and the Regional Internet Registries (RIRs).'[25]

India and IG at a Global Level

Arun Mohan Sukumar, in an article in *The Hindu*, has suggested that the Indian government must engage with the ICANN on four issues:

> first and foremost, the proposed handing over of the Internet Assigned Numbers Authority (IANA)—the specific department of ICANN responsible for running the DNS—from U.S. hands to a genuine, global organization; second, the role and continued relevance of ICANN's Governmental Advisory

---

[24] See Aaronson, S.A. 2012. 'Can Trade Policy Set Information Free?', *Vox*, available at https://www2.gwu.edu/~iiep/assets/docs/papers/2014WP/Aar onsonIIEPWP20149.pdf; accessed on 11 June 2018.

[25] Hill, R. 2014. 'Internet Governance: The Last Gasp of Colonialism, or Imperialism by Other Means?', in R. Radu, J.-H. Chenou, R.H. Weber (eds), *The Evolution of Global Internet Governance: Principles and Policies in the Making*, Berlin and Heidelberg: Springer-Verlag, p. 85.

Committee (GAC); third, the contractual relationship between ICANN and Verisign, Inc., a private Virginia-based company that operates as the Internet's 'registrar' in charge of the .com domain; and fourth, the security and resilience of India's .in and .bharat domains, as well as equal representation for all country code top-level domain operators at the negotiating table.[26]

For the moment though, it is clear that the Indian government's negotiations with ICANN have been limited, a reality that ICANN has publicly expressed regret about.[27] Nevertheless, and in the context of ICANN taking a lead role in formulating the IANA transition, the GoI, in its response, has clearly articulated its unease with ICANN's absence of external accountability and reliance on internal accountability, as well its abrogation of the power once enjoyed by the NTIA: 'The principle of external accountability is absent from the 2nd Draft Proposal, since ICANN will become the contracting authority for the naming function the sole venue for decisions relating to naming policy ... There would be no external checks and balances against the powers to be exercised by ICANN.'[28]

The Indian position is caught between MSH, which is the dominant model favoured by the Organisation for Economic Co-operation and Development (OECD) countries, and multilateralism that the Indian position gravitates towards, although this position is also favoured by authoritarian states such as Iran and China. In India, the major NGOs involved in IG-related issues, including IT for Change and the CIS, along with all the major information–trade bodies, including NASSCOM and

[26] Sukumar, A.M. 2014. 'Why India Must Engage with ICANN', *The Hindu*, 5 December, available at http://www.thehindu.com/opinion/op-ed/why-india-must-engage-with-icann/article6662319.ece#comments; accessed on 10 June 2018.

[27] Alawadhi, N. 2015. 'India Could Lose Opportunity to Have Greater Say in the Way INTERNET is Governed Worldwide: Fadi Chehadé, ICANN', *The Economic Times*, 10 March, available at http://articles.economictimes.indiatimes.com/2015-03-10/news/59970002_1_fadi-chehade-icann-internet-governance; accessed on 2 June 2018.

[28] 'Comments of Government of India on the *2nd Draft Proposal* of the Cross Community Working Group to Develop an IANA Stewardship Transition Proposal on Naming Related Functions', 2015, available at https://community.icann.org/pages/viewpage.action?pageId=53776547; accessed on 11 June 2018.

Federation of Indian Chambers of Commerce and Industry (FICCI), are divided on the matter of the specific governance architecture of IG. IT for Change, for example, is open to ICANN's multistakeholder model in the matter of 'technical' standards, but it argues for a far greater involvement of the State in the matter of the public provisioning of the Internet and therefore, in its governance.

The following account of the various stances on IG that India has taken over the last five years does seem to indicate a confusion and inability to articulate a clear position and/or encourage the BRICS nations to take a unified position. Part of the reason for the lack of a clear position is that there are clear pressures from both civil society and the private sector in the BRICS countries that favour a multistakeholder solution. There are internal pressures as well: for example, differences in opinion between key ministries within India, the MEA and the Ministry of Communications and Information Technology, that were expressed at NETmundial. At the India–Brazil–South Africa (IBSA) Conference held in Rio in 2011, all three countries proposed a new UN agency to deal with IG. At the UN General Assembly held in October 2011, the then opposition party spokesperson had suggested a 50-nation UN CIRP. In 2012, at the World Conference on International Telecommunication (WCIT) organized by the ITU in Dubai, where the key agenda item was to update the International Telecommunications Regulations (ITRs) that govern telecommunication network practices throughout the world, India, under pressure from its NGO lobbies, refused to sign the treaty along with the USA and other countries. The Indian position at the WCIT was specifically opposed to the Internet Resolution (Resolution Five). However, in a speech by an Indian official at NETmundial in April 2014: 'the term "equinet" was used to denote the quality being ascribed to the purpose of the internet. These competing philosophies, while moving towards the aim of keeping the internet "free" and a "global common good", have often put India at odds with the US, Europe and other countries at global internet governance meets.'[29] At the ICANN meeting held in London in June 2014, the Indian delegation consisted

---

[29] Kaul, M. 2014. 'Lessons from BRICS: Developing an Indian Strategy on Global Internet Governance', Science, Technology & Security Forum, available at http://stsfor.org/content/lessons-brics-developing-indian-strategy-global-internet-governance; accessed on 2 June 2018.

of relatively low-key staff, signalling the country's position on the role of ICANN.[30] With the present government's interest in positioning India as a key driver of the digital economy, India's position on IG at international meetings, including the ITU Plenipotentiary in November 2014 and the UN Commission on Science and Technology for Development (CSTD) Working Group on Enhanced Cooperation meetings held in late 2014 and early 2015, was robust, although it highlighted the levels of confusion related to IG.[31] At the Plenipotentiary Conference of the ITU held in November 2014, the Indian delegation pushed for a fair and equitable distribution and allocation of Internet Protocol addresses—for Internet Protocol version 6 (IPv6) address blocks to be allocated to countries— and a role for ITU in this process. The Indian government put forward a draft resolution on ITU's role in realizing a secure information society and recommended the ITU's Telecommunications Standardization Bureau:

a) to explore the development of naming and numbering system from which the naming and numbering of different countries are easily discernible;
b) to develop principles for allocation, assignment and management of IP resources including naming, numbering and addressing which is systematic, equitable, fair, just, democratic and transparent;
c) to make recommendations on network capability which ensure effectively that address resolution for the traffic originating and intended to be terminated by the user in the same country/region takes place within the country/region.[32]

[30] Kaul, M. 2014. 'Global Internet Governance: India' Search for a New Paradigm', Observer Research Foundation Issue Brief No. 74, August, New Delhi, pp. 1–12, available at http://www.globalpolicyjournal.com/sites/default/files/ORF%20Issue%20Brief%2074%20Mahima%20Kaul_0.pdf; accessed on 3 June 2018.

[31] Kovacs, A. 2014. 'Is a Reconciliation of Multistakeholderism and Multilateralism in Internet Governance Possible? India at NETmundial', Internet Democracy Project, 4 September, available at http://internetdemocracy.in/reports/india-at-netmundial/; accessed on 11 June 2018.

[32] 'India (Republic of) Proposals for the Work of the Conference—Draft New Resolution: ITU's Role in Improving Networking Functionalities for Evincing Trust and Confidence in IP Based Telecom Networks', Document 98 (Rev.1) E, Plenipotentiary Conference (PP14), Busan, 2 November 2014.

The issues raised by the Indian delegation are of critical importance given that at present and under the current regime, IANA allocates these addresses to five regional Internet registries who distribute these addresses to local or national registries.[33] The Indian government has argued that in the context of IPv6 and the fact that India will become a leading source for Internet traffic in the near future, the management of these addresses locally will make for a more efficient Internet. A long-term observer of Internet politics in India, Anja Kovacs from the Internet Democracy Project, has argued that India's position on IG remains ambivalent even as it presents a more unified perspective in the run-up to WSIS+10:

> Seeing the multitude of alliances that India continues to invest its energies in, it is unlikely, however, that India will want to let old allies down by breaking ranks—whether it is with BRICS, the Shanghai Cooperation Organisation (SCO), or the G77. Indeed, it is quite likely that India's embrace of multistakeholderism while re-emphasising the pre-eminence of governments when it comes to cybersecurity issues is a predominantly tactical move: designed to first and foremost expand the number of allies it has, rather than indicating a dramatic policy shift with far-reaching practical implications.[34]

Any evaluation of the Indian government's response to IG needs to deal with its historical ambivalence towards the multistakeholder model, especially at the federal level where NGOs and special interest groups have only rarely been involved as full participants in national policymaking. There have been exceptions, such as the RTI Movement in which key activists such as Aruna Roy and Nikhil Dey were involved and the role played by NGOs in discussions leading up to the amendment of the Copyright Act in 2013 that gave people with visual impairment access rights. India has consistently taken the position that the governance of the Internet should be dealt with by sovereign states. While the Indian government's ambivalence has invited ire from the advocates of the multistakeholder status quo, it is clear that this status quo is untenable

---

[33] Saran, S. 2014. 'The ITU and Unbundling Internet Governance: The Indian Perspective', Council on Foreign Relations, available at http://www.cfr.org/internet-policy/itu-unbundling-internet-governance/p33656; accessed on 4 June 2018.

[34] Kovacs, A. 2015. 'India', pp. 78–81, available at http://wsis10.asia/documents/04-WSIS+10-India-150828.pdf; accessed on 8 June 2018.

given that it did, through omission or commission, contribute to the surveillance regime uncovered in the Snowden revelations. The far greater issue is the continuing influence of the USA in policymaking related to the Internet. Reporting from Busan on the ITU plenipotentiary, Samantha Dickinson, writing in *The Guardian*, highlights the nature of realpolitik and the consequential lack of progress on the matter of IG:

> back room negotiations spearheaded by the US delegation meant many of the changes proposed by other countries were taken off the table. Those negotiations took placed behind closed doors, but it is understood that the US gave up its demand to have non-governmental groups invited into ITU's council working groups, which were designed to be for governments only. In return, other states withdrew proposals about online privacy, cybersecurity and other internet proposals. No major threats to the internet have emerged as a result of the conference. Instead, many of the hottest internet issues have been shunted off to a small group of the ITU, known by the convoluted name of the Council Working Group on International Internet-related Public Policy Issues, or CWG-Internet for short.[35]

The present government's Digital India initiative is yet to deal with the issue of IG, although there is evidence that the Prime Minister's Office is increasingly becoming involved in IG-related matters. It is, however, abundantly clear that the country will need to clarify its position on IG and support policies that expand the potential of the Internet as a tool for equitable and justice-based growth. Part of the reason for this confusion has been the turf wars between the MEA that have taken a traditional line related to Indian sovereignty over the Internet and DeitY that supports business and the IT industry and goes with the flow on the issue of MSH. If the Indian government is neither keen on MSH or multilateralism, it must articulate a position that is inclusive, supports diversity and a multiplicity of interests, from a position that recognizes that all of India's citizens have a stake in the future of the Internet in India. The GoI did clarify its support for multistakeholder solutions at the ICANN 53 held in Buenos Aires in June 2015, though it refrained

---

[35] Dickinson, S. 2014. 'How will Internet Governance Change after the ITU?', *The Guardian*, 7 November, available at http://www.theguardian.com/technology/2014/nov/07/how-will-internet-governance-change-after-the-itu-conference; accessed on 4 June 2018.

from specifying its preferred model of MSH. This turn towards MSH gives us an opportunity to reflect on the geopolitics of this decision. For example, it could signify India's closer cooperation with the US on matters of international and national cybersecurity that is high on the agenda of US–India relationships.

The move towards MSH, also preferred by the Brazilian government, has compromised any BRICS-based solutions given that both Russia and China continue to take a multilateral position on IG. The decision to embrace MSH could have been based on very pragmatic reasoning given that India's alliance with the US on the matter of curbing the influence of China in the region is increasingly a priority foreign policy issue. Compromising on IG will be seen as a small price to pay for greater cooperation and the strengthening of trade links in a host of other areas. Earlier this year, there was a report in the Indian media regarding India negotiating the siting of a 'root server' in India in return for compromising on the issue related to the modalities of IG (at present, there are 13 root servers: 10 in the USA, 2 in Europe, and 1 in Japan).[36] If they do follow this line, then it would be close to impossible for countries such as India to contribute to the reformation of IG based on a sharing of power and involvement in taking decisions on the 'architecture' of the Internet, placing restrictions on snooping, and unilateral policymaking on the Internet. However, given the strongly centralist tendencies of the present government, the fact that India was silent on the specific contours of MSH at ICANN 53 in Buenos Aires does suggest that it might place its weight behind efforts to push for a greater say for 'sovereign states' in any new MSH model related to IG. That position is clearly evident in an article in *The Indian Express* post the IANA transition:

> While ICANN will now be governed by a 'multi-stakeholder' model, including businesses, individual users, India's push for a multi-stakeholder model envisages a pivotal role for governments as the custodian of cyberspace in the wake of security threats from terror groups. India has described the role of the government as 'an important stakeholder' and 'a custodian of security' for the

[36] Samanta, P.D. 2015. 'Internet Governance: US Considering India's Pitch to Locate "Root Server"', *The Economic Times*, 13 September, available at http://articles.economictimes.indiatimes.com/2015-09-03/news/66178452_1_internet-governance-root-thirdlargest-internet-user-base; accessed on 3 June 2018.

global Internet infrastructure. India's proposal, as enunciated in Marrakesh, is that the Internet should be managed through the multi-stakeholder approach, and that governments should have 'supreme right and control' on matters relating to international security. India in its submission has said that under the new transition, the body managing the Internet should have 'accountability towards governments' in areas where 'governments have primary responsibility, such as security and similar public policy concerns'.[37]

The fact that ICANN's (an organization that is registered under California law) overseeing of IANA's responsibilities has been contested by US senators such as Ted Cruz does imply that the transition will not result in the relinquishing of ultimate US oversight, except in a symbolic sense—a view that was aired in a recent US Senate hearing on the IANA transition.[38] It is also unclear whether hosting the first ICANN meeting (ICANN 57) in Hyderabad, after the US Department of Commerce gave up its oversight, between 3 and 6 November 2016, has paid any dividends to the GoI. There were three key issues for the GoI: (i) equal ownership and access to TLD names; (ii) involvement in the making of technical standards for non-English languages; and (iii) creating capacity in Domain Name System (DNS) expertise.[39]

The State and IG

There is an assumption in the literature on IG that in an era of MSH, the role of the State in the governance of the Internet ought to be limited.

[37] Sasi, A. 2016. 'Soon, the Net Will be Free of US Control, Have New Governors. In New ICANN, Who Can?', *The Indian Express*, 21 March, available at http://indianexpress.com/article/explained/soon-the-net-will-be-free-of-us-control-have-new-governors-in-new-icann-who-can/; accessed on 1 June 2018.

[38] 'US Senate Holds Hearing on IANA Stewardship Transition and Its Impact on "Internet freedom"', 2016. Mayer Brown, 21 September, available at https://www.mayerbrown.com/files/Publication/5ad74332-1269-4da2-974a-68923988d747/Presentation/PublicationAttachment/f94046ff-e0bd-4fc3-b098-728cd1e31704/160922-UPDATE-Internet.pdf; accessed on 11 June 2018.

[39] Srivas, A. 2016. 'Modi's Internet Diplomacy Gambit Marks Symbolic Milestone with ICANN Summit at Hyderabad', *The Wire*, 4 November, available at https://thewire.in/77970/internet-governance-india-icann-hyderabad/; accessed on 27 June 2018.

There are of course exceptions, such as the USA that uses its superpower status to control the terms of negotiations in both multistakeholder and multilateral forums. The multistakeholder model is seen as an advance over multilateral models that are state-centric, best illustrated by the UN system. The fact that the involvement of the State sector in ICANN is contained within its GAC does suggest that the influence of the State is proscribed within a system of governance that provides equal space to both the business and civil society sectors. The GAC exists to provide advice to ICANN on public policy issues; and its strictly advisory capacity does suggest that its power to influence ICANN is limited. But is that necessarily the case given that the State does have the power to shape governance at a local level that can influence the ways in which governance at a global level pans out? After all, it was the State sector that urged for changes to IG in the aftermath of the Snowden revelations. It is instructive that the Montevideo Statement (7 October 2013), issued in the aftermath of Snowden revelations by key IG actors, including the IAB, IETF, and W3C, among others, supported the status quo and was against the Balkanization of the Internet.[40] While the ccTLDs are currently managed by both State and non-state actors, as a public good and resource, there not much that ICANN can do if a government decides to wrest control of its ccTLD from a non-state actor. Park cites an example of ICANN's helplessness in the migration of control over a ccTLD from the non-state sector to the government:

> If a government seems determined to take over its ccTLD ignoring ICANN's authority as global coordinator, ICANN accepts the reality. In reality, ICANN has no legally binding power when it comes to ccTLD coordination. When the government of South Africa took over its ccTLD, ICANN could just observe. Initially, the .ZA ccTLD manager thought the ICANN community would have backed him up. It never happened.[41]

---

[40] 'Montevideo Statement on the Future of Internet Cooperation' 2013. 7 October, available at https://www.iab.org/documents/correspondence-reports-documents/2013-2/montevideo-statement-on-the-future-of-internet-cooperation/; accessed on 25 June 2018.

[41] Park, Y.J. 2009. 'The National CCTDL Disputes: Between State Actors and Non-State Actors', *International Journal of Communications Law & Policy* 13(4): 193.

In the context of India where the contribution of services to the GDP has increased exponentially over the years and the country has stated ambitions to become a knowledge power, it is to be expected that they will invest in core IG concerns at a national level, including control over its ccTLD (.in) and substantive control over cybersecurity. Van Eeten and Mueller, in an article that highlights the myopic view of IG that limits our understanding of IG to processes related to ICANN, WSIS, and IGF, point to the need to both factor in the role of the State and 'innovative areas such as the economics of cybersecurity, network neutrality, content filtering and regulation, copyright policing and file sharing, and interconnection arrangements' that are examples of IG in action.[42]

One of the issues then, in the context of a country such as India that has traditionally been involved in myriad ways in the provisioning of 'welfare', is whether the State should keep its distance from the provisioning of the Internet as a public resource. The multibillion-dollar Digital India project launched by the GoI in 2015 would suggest otherwise, for it is clear that the State is keen to play a key role in expanding access and resourcing the 'public' in India. Indian civil society is divided on the issue of State involvement in IG, with some groups such as IT for Change arguing that the State simply has to play a key role in IG precisely because only they have the mandate, resources, and will to expand Internet access for all in India. Other groups, such as the CIS, who are committed to Internet freedom have less to say about the role of the State or issues related to IG in the context of a country that is faced with major divides, including poverty, though they have lobbied for access for special sectors such as the visually impaired. The Chennai-based Internet Society of India seems to be a strong supporter of MSH and the ICANN status quo with respect to the involvement of the State in IG. What is clear is that unlike other resourcing issues that have attracted strong positions and lobbying from NGOs and academics aligned to the left in India, issues related to IG are yet to attract such focused attention and, in its absence, the views related to IG are motivated by liberal and libertarian views. This divide within civil society on issues related to the Internet is a reflection of the diversity of interests and claims on the Internet from different sections in society. Typically, civil society involved in education and health on behalf of

---

[42] Van Eeten and Mueller. 2012. 'Where Is the Governance in Internet Governance', p. 730.

poorer sections of society has formed clear policy positions vis-à-vis State policy. That is not the case with civil society involved in the Internet given that they cater to different interests—for example, the middle classes whose concerns related to the Internet have little to do with issues such as 'access' or 'affordable use' of the Internet but are typically from a consumer perspective. Parminder Jeet from IT for Change rues the fact that Internet think tanks in India now account for a large part of Foundation support, resulting in a situation where there is little recognition or funding for basic issues such as access in the context of poverty:

> Traditionally in India CS fought for the cause of the poor, the marginalized. They were slightly left of centre. Internet-related CS includes the full political spectrum from quite liberal to right of centre—very few that are left of centre. There are issues that affect richer people, consumer issues how Facebook is treating me—that is very different from how Facebooks affects redistribution. There are lots of liberal CS groups related to the internet. They take all the attention and the funding—even Foundations that typically fund CS on poverty alleviation—on the Internet, they support liberal CS—fund them to such an extent that very little is available for the framing of issues related to the Internet that are of greatest concern to the poor people—that framing is missing.[43]

Governance of Internet in India

Internet governance in India is shared between the State, quasi-governmental agencies, the private sector, and civil society, although the role of the State remains preponderant. In this sense, both the private sector and civil society either function in their capacity at the behest of the government, offering services that otherwise are not offered by the State, or provide advice to the State on a case-by-case basis. While key civil society groups in India, such as the CIS and IT for Change, do participate in State-run IG activities, their remit is not exhausted with their involvement with the State. Broadly speaking, the State's IG functions are carried out under the DeitY. The DeitY carries out a number of activities related to IG, including the following:

---

[43] Personal interview with Parminder Jeet Singh, IT for Change, 2 April 2015.

- Policy matters relating to Information Technology, Electronics and Internet.
- Promotion of IT and IT enabled services and Internet.
- Assistance to other departments in the promotion of E-Governance, E-Infrastructure, E-Medicine, E-Commerce, etc.
- Promotion of Information Technology education and Information Technology-based education.
- Matters relating to Cyber Laws, administration of the Information Technology Act. 2000 (21 of 2000) and other IT related laws.
- Interaction in IT related matters with International agencies and bodies.
- Initiative on bridging the Digital Divide, Matters relating to Media Lab Asia.

Promotion of Standardization, Testing and Quality in IT and standardization of procedure for IT application and Tasks.[44]

The DeitY is involved in expanding involvement in IG both at the domestic and international levels: R&D related to the move to IPv6; establishing a multilingual domain name system for India based on country code domain names in seven Indian languages; exploring access for the visually impaired through initiatives such as an open source Web browser; the development of infrastructure such as the National Internet Exchange of India (NIXI) and the .IN Internet Domain Registry; organizing awareness workshops; and being involved in GAC and the IGF. It has been involved in the Multistakeholder Advisory Group and the India Internet Governance Forum (IIGF) that was initiated by FICCI and that provides the space for multistakeholder dialogue on issues related to IG.[45] The IIGF consists of 39 members: 8 representing the government; 12 representing the private sector, particularly apex bodies such as NASSCOM, FICCI, and others; 7 members representing civil society, including the CIS and IT for Change; and 7 representing academia drawn mainly from the Indian Institutes of Technology (IITs) and Indian Institutes of Management (IIMs), along with 5 members representing

---

[44] DeitY, 'Business Rules', available at http://deity.gov.in/content/dit-business-rules; accessed on 11 June 2018.

[45] 'India Internet Governance Forum (IIGF): Constitution of Multi-stakeholder Advisory Group (MAG)'2014. F.No.L-13014/3/2013-Int.Gov., DeitY, 4 February, available at http://cis-india.org/internet-governance/blog/mag-order.pdf; accessed on 10 June 2018.

the technical community drawn wholly from government departments, including NIXI and C-DAC. This group's terms of reference include the review of the IG policy landscape, acting as a platform for discussions, and the provision of consensus-based positions to the Inter-Ministerial Group on matters related to IG. However, bodies such as the IIGF have not had support from the MEA. Arun Sukumar from the Centre for Communication Governance at the National Law University, New Delhi, said in a personal interview:

> The India internet governance forum hasn't really taken off since the MEA opposed the creation of IIGF because it was a FICCI-led initiative. It seems unlikely that it will take off under the FICCI umbrella but perhaps it could manifest itself in some other way. There is also a multistakeholder advisory group. It ran its course once. It has been reconstituted but as far as I know it has not met after its reconstitution.[46]

While the constitution of the IIGF is certainly an important achievement considering that the Indian government has traditionally had a frosty relationship with its NGOs, its modalities of participation, engagement, discussion, and debate are still in its infancy and there are as yet no take-away learnings from this experience, especially for civil society. The fact that the IIGF has not met for close to a year does, however, suggest that the BJP-led government is yet to configure a clear approach to IG. The Indian government's position has veered from a multilateral stance to a multistakeholder one—positions that are also reflected in civil society organizations that have only episodically been involved in matters related to IG. The private sector too has had limited involvement in matters related to IG.[47] While large countries like India and China do want to have a more substantive role at ICANN especially in the light of the apparent ceding of control over ICANN by US-based entities in 2016, India's 'schizophrenic' approach to IG is illustrated by its support, at one time or the other, for four different models of IG highlighted by Subramanian: 'The ITU (International Telecommunications Union)-

---

[46] Personal interview with Arun Sukumar, 9 April 2015.

[47] Alawadhi, N. 2016. 'India Can Have Bigger Say with ICANN Managing the Internet', *The Economic Times*, 4 October, available at http://economictimes. indiatimes.com/tech/internet/india-can-have-bigger-say-with-icann-managing-internet/articleshow/54664943.cms; accessed on 7 June 2017.

based *inter-governmental model*; the IGF (Internet Governance Forum)-based *inter-governmental + equal multi-stakeholder model*; the UN-CIRP (Committee on Internet-related Policies) based *inter-governmental + limited stake-holder model*; and the *fully participatory model* where the ICANN becomes a completely independent body without any supervision from any government or UN Agency (emphasis in original).[48] India, in other words, has not been able to successfully bring the Internet as infrastructure under its control.

One of the clear examples of the State's involvement in IG is reflected in its clear stance on the development of ccTLDs and Uniform Resource Locators (URLs) in local languages beginning with the official language, Hindi. While English, for all practical purposes, is the lingua franca of the educated middle classes in India, Hindi does have a national imprint and is spoken by more than half of India's population. Language barriers to access and identity remain an issue with the Internet. The need for local language domain names was initiated by the government at ICANN in 2010. Since the Devanagari script is used by 11 official languages in India, the first internationalized domain name (IDN) will be .भारत (dot Bharat),[49] along with related extensions. The .IN Registry under DeitY and the Internet Service Providers Association of India (ISPAI) will be in charge of rolling out this ccTLD. This is the beginning of the indigenization of the Internet in India that will expand access and use of the Internet across India.

In other words, this example clearly shows that the Indian state is not a passive observer in the matter of IG but is actively involved in shaping the architecture and modalities of the Internet in India, while also contributing to governance at an international level through GAC and other bodies. While the Indian state is, as per its culture, bound to follow hierarchical processes of control, it does recognize the need to delegate control of some aspects of IG to the private sector, especially organizations that function under the aegis of established trade associations, such as

---

[48] Subramanian, R. 2013. 'Internet Governance: A Developing Country Perspective', *Communications of the IIMA* 13(4): 12.

[49] ':भारत (.BHARAT) Country Code Top Level Domain (ccTLD) Name: Internationalized Domain Names (IDNs)—.IN Domain Registry', available at http://cdac.in/index.aspx?id=pdf_IDN_policy_framework; accessed on 11 June 2018.

NASSCOM and ASSOCHAM. A good example of a private–public initiative related to IG is the role played by the Data Security Council of India (DSCI), an initiative of the NASSCOM that nevertheless is closely linked to DeitY, the Department of Commerce, Ministry of Home Affairs, National Security Council Secretariat, Planning Commission, the Central Bureau of Investigation, and the Bureau of Indian Standards. The DSCI's key role is to train law enforcement personnel to handle and respond to cybercrimes. It is an independent self-regulatory organization whose mandate is linked to the business of data protection. The DSCI is also involved in contributing towards IG in areas such as trans-border data flows, cybersecurity, privacy, encryption, assisting the government on policy on cloud computing and technical standardization, and is represented on the IIGF. Similarly, the .IN Registry has been set up by the NIXI and is promoted by the DeitY in association with ISPAI, the key apex body for ISPs in India. The experience of India reflects, to some extent, the experience of states in Europe that have adopted a mixed approach to the management of ccTLDs (albeit with a stronger role for the State). Christou and Simpson, in an article that explores the patterns of ccTLD governance in Europe, conclude by observing that:

> ccTLDs provide an interesting example of a novel form of governance in which the presence of the state has been far from abandoned. Rather, in Europe, the state sits within a network of decentralised systems that, in operational and managerial terms, vary in their degree of hierarchy. This is distinctly different from merely creating a free market governed by competition law. Instead the ccTLD registry plays the key governance role in the functioning, but also the policy evolution, of its TLD. The system relies on a balance between the functional dynamism and managerial efficacy of the registry, on the one hand, and its willingness to listen to the advice given by the pluri-interest characterised advisory boards that often monitor sectoral activity, on the other.[50]

The Pitfalls of PPPs: The Case of Cybersecurity

While there are some areas related to IG that from a PPP perspective can be considered relatively unproblematic, such as initiatives related to the

[50] Christou, G. and S. Simpson. 2009. 'New Governance, the Internet, and Country Code Top-Level Domains in Europe', *Governance: An International Journal of Policy, Administration and Institutions* 22(4): 620–1.

expansion of Internet access, the indigenization of domain names, and even engineering solutions that expand public access, there are a number of areas in which such partnerships may impact negatively on the public interest. These include privacy issues in the context of PPPs related to cybersecurity, surveillance, and access to private data by private companies involved in large, State-sponsored e-government projects, as well as the interception of private data by firms that are, by law, required to turn this data to State security agencies. While PPPs are most certainly required for scaling up large-scale access projects in India, the issue of whether such initiatives enhance 'risk' remains a critical issue.

Ulrich Beck, who pioneered studies on the risk society, suggests the need for a 'relations of definitions'; in other words, clarity related to the role played by 'rules, institutions and resources' in the mitigation of risks such as environmental disaster. The following four questions can be applied to understanding risk not only in terms cybersecurity but also e-governance and 'development' projects supported by the State and the private sector in the context of contemporary India:

(1) Who-that is, what social agency and authority establishes in what way how harmless or dangerous products and their side-effects are? Does the responsibility lie with those who create and profit from the risks, or with those who are currently or potentially affected, or with public agencies? (2) What type of knowledge or unawareness of causes, dimensions, agents and so on is consulted or acknowledged here? Who bears the burden of proof? (3) What is considered 'sufficient proof' and this, of course, must be answered in a world where all knowledge of hazards and risk moves in the presuppositions of probability theory. (4) Where hazards and destruction are recognised and acknowledged, who decides issues of liability, compensation and costs for the affected parties, and who rules on appropriate forms of future monitoring and regulation?[51]

Beck thus highlights the need to mitigate the consequences of organized and unorganized irresponsibility: for example, the witting and unwitting culpability of large software firms, such as Microsoft, who are involved in e-governance in India and who, therefore, have access to vast amounts of data on the public. Another example of such risk is sensitive private data falling into the hands of companies involved in the UIDAI scheme. While

---

[51] Beck, U. 1997. 'Global Risk Politics', *The Political Quarterly* 68(B): 29.

it is clear that 'self-regulation' is the preferred model for companies such as the DSCI, the question that can be raised is whether there is need for more rather than lesser oversight by the State over such organizations given their access to critical information on individuals and their behaviours on the Internet.

In the case of surveillance, there has been an extraordinary increase in both PPPs and firms involved in supplying the State with surveillance hardware and software.[52] A report on the surveillance industry in India on the CIS website (19 February 2015) highlights 13 surveillance expositions in India, along with the key exhibitors, visitors, and technologies on show at these events.[53] The technologies, as per the report, included video surveillance and analysis, biometric identification, mass data gathering and analysis, cell-phone location tracking and vehicle monitoring, and air/ground drones and satellite surveillance. Under data gathering and monitoring, the following, mainly non-Indian companies, were identified:

Cobham, Comguard, Cyint, ELT (UK), Fastech, Hacking Team (Italy), Smoothwall (USA), Verint Systems (USA), Cyint technologies, Atlas Electronik (Germany), Audiotel International (UK), Avancar, Cobham (UK), ELT (UK), Eyewatch, Kommlabs, Mangal Security Systems, Merit Lilin (Taiwan), Ockham Solutions (France), Septier (Israel), Synway (China), ACSG Corporate, Amesys (France), Anritsu (Japan), Axis (Sweden), BAE Systems (UK), Blue Coat (USA), C-dot, Comint, Cyberoam (USA), Deviser Electronics, Elsira (Elbit) (Israel), Esri (USA), Exelis, General Dynamics (USA), Helyx (UK), ITP Novex (Israel), Leica (Switzerland), Net Optics (Ixia) (USA), Northrop Gruman (USA), Rahul Commerce, Rohde And Schwarz (Germany), RVG Diginet, Tas-Agt, Trueposition (USA), Zte Technologies (China).[54]

The situation in terms of cybersecurity is equally if not more problematic, given that agencies such as the DSCI act on behalf of private bodies such

[52] Xynou, M. 2014. 'The Surveillance Industry in India', CIS, March, pp. 1–48, available at http://cis-india.org/internet-governance/blog/surveillance-industry-india.pdf; accessed on 23 June 2018.

[53] Joshi, D. 2015. 'The Surveillance Industry in India: An Analysis of Indian Security Expos', 19 February 2015, available at http://cis-india.org/internet-governance/blog/surveillance-industry-in-india-analysis-of-indian-security-expos; accessed on 2 June 2018.

[54] Joshi, 'The Surveillance Industry in India', pp. 1–48.

as NASSCOM to fulfil State interests. In addition to the State-based Indian Computer Emergency Response Team (CERT-In), the National Security Council Secretariat, the National Cyber Coordination Centre, the Operational Group on Cyber Security, the National Informatics Centre, the National Information Security Assurance Programme, National Critical Information Infrastructure Protection Centre, and the CMS are among others involved in combating threats to the State from hacking, trojans, computer viruses, email-related crimes, and so on. Their actions are mandated by the National Cyber Security Policy of India, 2013, a joint working group that was established with the private sector in 2012. The State is currently exploring a variety of PPPs based on service and lease contracts, concessions, and joint ventures. There are private agencies and firms, such as Mirox Cyber Security & Technology Pvt Ltd, the Information Sharing and Analysis Centre, Perry4Law Techno Legal Base (PTLB), and Cyber Safe India, which have been contracted by the State to manage specific initiatives related to cybersecurity. The National Security Database is a private initiative whose objective is to create a database of 'ethical hackers' and cybersecurity personnel who can be called upon by the State in the context of a cyber emergency. Cyber Safe India's partners include the Ministry of Information Technology, the National Cyber Security Alliance, Computer Society of India, and Internet Security Society.[55] Growth in the cyber-security sector was to some extent hastened by the Mumbai terror attacks in 2008, although it would seem that many of the private firms have been established to take advantage of the availability of public funds to combat cyber security. While the government has demanded access to all data flows that are part of the normal business of companies such as Blackberry, it is yet to invest large amounts of money in encryption technologies. There is, so far, little evidence that such partnerships have contributed to enhancing cybersecurity in India. On the contrary, there are major questions related to the limitations to the freedom of expression, the right to privacy, and the right to opinions in cyberspace as the State, via clauses such as 66A in the IT Act, had arbitrarily clamped down on anyone expressing dissident opinions, including critiquing ministers. On 25 March 2015, the Supreme Court scrapped 66A stating that it was unconstitutional and

---

[55] See Joshi, 'The Surveillance Industry in India'.

compromised every citizen's freedom to express opinion. Sinha, writing in the *Hindustan Times*, explains the verdict:

> The top court held Section 66A was vague and didn't explain what content should be construed as 'annoying', 'inconvenient', 'grossly offensive' or of a 'menacing character', making it open to abuse. The section made no distinction between mere discussion or advocacy of a particular point of view—which may be annoying or inconvenient or grossly offensive to some—and incitement, by which such words lead to an imminent causal connection with public disorder and security of the state.[56]

The foregoing discussion on PPPs highlights a significant flaw in the multistakeholder model adopted by the Indian government in the matter of IG. While the State has allied itself with a number of NGO initiatives aimed at strengthening cybersecurity in India, it has not aligned itself with those groups that are in the forefront of advocating for citizen's digital rights in the context of a situation in which security has become a fundamental plank of the State's cyber policy; also, equally important areas such as privacy have been ignored. While groups advocating for such rights—such as the CIS—do take part in the IIGF, the State's agenda on national security is the privileged discourse and discursive bodies that adopt a contrary position are at best tolerated in the spirit of MSH. Moreover, and in the light of the Snowden revelations, its continuing involvement with the India–US ICT Working Group,[57] which met in January 2015 in Washington to discuss cross-border data flows, cybersecurity, IG, and e-commerce, perhaps makes sense from the point of view of trade, but there remain large issues related to snooping that are as yet unresolved.

One of the challenges for the State in India in the context of IG has been to operationalize communicative actions based on multistakeholder

---

[56] Sinha, B. 2015. 'Supreme Court Upholds Free Speech on Internet, Scraps "unconstitutional" Section 66A of IT Act', *Hindustan Times*, 25 March, available at http://www.hindustantimes.com/india-news/supreme-court-uphold-free-speech-online-strikes-down-vague-section-66a-of-it-act/article1-1329903.aspx; accessed on 30 June 2018.

[57] 'Joint Statement of the US–India Information and Communication Technology Working Group Meeting', 2015. US Department of State, Washington, 14–15 January, available at http://www.state.gov/r/pa/prs/ps/2015/01/236080.htm; accessed on 11 June 2018.

negotiations.[58] The Indian state has traditionally not dealt with civil society in the making of development policy and is not known for its support for an agonistic public sphere. If IG is all about 'the development and application by governments, the private sector, and civil society, in their respective roles, of shared principles, norms, rules, decision-making procedures, and programs that shape the evolution and use of the Internet,'[59] then it is clear that the State in India is yet to fully operationalize this process. In other words, one can argue that the Indian state's governance remains hierarchical, despite the fact that it has initiated the IIGF on the basis of the multistakeholder model of IG.

While the US government's unwillingness to relinquish its hold over critical IG infrastructure is becoming clearer by the day, it would seem that there has been a toning down of the criticism of the US oversight from countries such as Brazil and India and a willingness to explore MSH solutions. In other words, any radical overhaul of IG seems to be a distant possibility unless countries such as Russia and China launch their own solutions to IG. For the moment, however, it seems likely that a slightly reformed IG status quo is all that is on offer. Given the present government's close trade and foreign policy ties to the USA, it is highly unlikely that it will take a radical stance on IG. However, if indeed it is to embrace MSH solutions, it must operationalize it locally. In the words of Sukumar:

> The Indian government's ringing endorsement of multistakeholderism stands in contrast to its own top-down diktats on key internet concerns, whether it is net neutrality, freedom of speech and expression on the web, or the liability of online intermediaries. If New Delhi is indeed convinced of the merit of bringing all stakeholders—including civil society and technical experts—on board, it must break the cosy relationship between Indian industry and regulators in the ICT sector. To lend credibility to India's interventions abroad, the government must walk its talk on multi-stakeholderism at home.[60]

---

[58] Risse, T. 2005. 'Global Governance and Communicative Action', in David Held and Mathias Koenig-Archibugi (eds), *Global Governance and Public Accountability*, Malden, Oxford, and Carlton: Blackwell, pp. 164–289.

[59] 'Report of the Working Group on Internet Governance', Chateau de Bossey, June 2005, p. 4, available at https://www.wgig.org/docs/WGIGREPORT.pdf; accessed on 26 March 2019.

[60] Sukumar, A.M. 2015. 'India's New "Multistakeholder" Line Could be a Gamechanger in Global Cyberpolitics', *The Wire*, 22 June, available at http://

Sukumar's critique is important precisely because in the context of the UN General Assembly High Level Meeting on WSIS+10 Review, on 15 December 2015, the secretary of DeitY, Mr J.S. Deepak, reiterated India's support for MSH and the role of stakeholders, although he insisted for a more expanded role for the State in specific matters related to IG: 'In the context of security and allied public policy concerns, we believe that governments, which bear ultimate responsibility for essential services and for public safety, have a key role to play and be central to discussions regarding security of the internet.'[61] In other words, it is clear that India, along with other countries such as China, will continue to push for the 'strategic autonomy' of the State in critical areas related to IG, such as cybersecurity, and will continue to play a dominant role in setting up the modalities of MSH participation and the terms of involvement for civil society at a local level. Anuj Srivas, writing in the immediate aftermath of ICANN 57 held in Hyderabad, India, in early November 2016, states that for all of India's public commitments to MSH, there has been little progress reflected in Bills such as the Aadhaar Unique Identity Bill that has been passed without any significant consultations with stakeholders.

On policy issues that would ideally require a multi-stakeholder touch, India is still struggling. There are examples of issues that currently have no multi-stakeholder process but would ideally require one, instances of where multi-stakeholder processes are carried out in initial stages but remain opaque and non-inclusive during later stages, technology-related concerns that kick off a multistakeholder process but ultimately wind down without proper conclusion and finally policy issues where only one or two voices and perspectives are taken into account in a particularly detrimental fashion.[62]

---

thewire.in/2015/06/22/indias-new-multistakeholder-line-could-be-a-gamechanger-in-global-cyberpolitics/; accessed on 22 June 2018.

[61] 'Statement by J.S. Deepak, Secretary, Department of Electronics & Information Technology at the United Nations General Assembly High Level Meeting on WSIS+10 Review', Permanent Mission of India to the UN, 15 December 2015, available at https://www.pminewyork.org/pages.php?id=2340; accessed on 4 June 2018.

[62] Srivas, A. 2016. 'Ravi Shankar Prasad Walks the Multistakeholder Line in Hyderabad, but It Doesn't Extend to Delhi', The Wire, 7 November, available at https://thewire.in/78229/india-multistakeholderism-icann57-meeting-internet/; accessed on 3 June 2018.

It is, however, interesting to note that despite State inaction on the matter of progressing MSH, there are initiatives such as the India School of Internet Governance (inSIG)[63] that is involved in training and building IG capacities in individuals representing government, business, civil society, technical, or academic interests.[64]

We do, however, need to place in perspective India's bids to negotiate the geopolitics of information within a context in which its chief protagonist and ally is the USA. India's ambitions with respect to the knowledge economy may be huge, but its exercise of power at a global level or geopolitical heft cannot be compared with that of China, whose geopolitical ambitions are reflected in its annexation of territory, the expansion of trade, aid, and investments, it becoming a key location for the generation of revenues for the world's leading cultural industries such as Hollywood, its massive domestic market, increase in connectivity, and moves to create its own, autonomous information sphere. It is China and not India that poses a threat to the USA's ambitions to remain at the helm of informational capitalism. In Dan Schiller's words: 'This new geopolitics of information ... is becoming more complex. It is likely that the U.S. will offer tactical concessions, however, U.S. political leaders are becoming more aggressive in asserting U.S. digital capital's international claims— notably, with respect to China.'[65] In other words, India's ambitions to attain 'strategic autonomy' in its 'informational' realm are bound to be fraught precisely because it is in a qualitatively different dependent relationship with the USA, unlike China. However, and despite all the posturing from the Indian government on MSH, the Chinese government's adoption of a new cybersecurity law on 1 July 2017, based on top-down processes, is probably where the GoI would really like to be on the matter of IG.[66]

[63] The schools were organized in Hyderabad in 2016, Trivandrum in 2017, and Kolkata in 2018 and 2019.

[64] See plans for InSIG 2019: 'India School on Internet Governance', 15–17 November, Kolkata, available at https://insig.in/; accessed on 26 March 2019.

[65] Schiller, D. 2015. 'Geopolitics and Economic Power in Today's Digital Capitalism', Paper presented at the Hans Crescent Seminar, London, 31 December, available at http://informationobservatory.info/2015/12/14/geopolitics-and-economic-power-in-todays-digital-capitalism/; accessed on 2 June 2018.

[66] Kleinwachter, W. 2017. 'Internet Governance Outlook 2017: Nationalistic Hierarchies vs. Multistakeholder Networks', CircleID, 6 January 2017, available at http://www.circleid.com/posts/20160106_internet_outlook_2017_nationalistic_hierarchies_multistakeholder/; accessed on 3 June 2018.

6

The WIPO Treaty for the Visually Impaired as a Double Movement

In the annals of advocacy related to global media governance, there are two outstanding case studies. First, the state-based mobilizations linked to the decolonization of information that was on the agenda of the Non-Aligned Movement in the 1960s and 1970s, and that culminated in United Nations Educational, Scientific and Cultural Organization's (UNESCO) call for a New World Information and Communication Order (NWICO). It led to the MacBride Commission's study of the 'world's communication problems' in 1979 and resulted in its recommendation to change power relations in global communications. The second were the two multistakeholder gatherings related to the WSIS, a UN-based event that was held in two phases—Geneva, 2003 and Tunis, 2005—under the aegis of the ITU. Both events have attracted sustained academic attention. While the history and politics of NWICO have been recounted by a number of academics, including Golding and Harris, Hamelink, and Nordenstreng and Schiller,[1] along with practitioners such as Righter,[2] the experience of the WSIS, in particular the role played by civil society, has been the focus of sustained academic attention by Mueller, Kuerbis,

---

[1] Golding, P. and P. Harris (eds). 1997. *Beyond Cultural Imperialism: Globalization, Communication and the New International Order*. London: SAGE; Hamelink, C.J. 1994. *The Politics of World Communication*. London, Thousand Oaks, and New Delhi: SAGE; and Nordenstreng, K. and H. Schiller (eds). 1979. *National Sovereignty and International Communications*. Norwood, NJ: Ablex.

[2] Righter, R. 1978. *Whose News?: Politics, the Press and the Third World*. London: Burnett Books.

and Page; O'Siochru; Raboy; and Thomas,[3] among a plethora of scholars. While NWICO's call for communication rights polarized the world into the communication haves and have-nots and led to the marginalization of UNESCO for its temerity to challenge media power, the WSIS's key achievements included the creation of IGF and providing an opportunity for civil society to mobilize, coordinate, campaign, and advocate for information rights at the level of the UN. While NWICO undoubtedly made the first case for global media reform, the WSIS too has been a watershed moment in the history of global media advocacy and it gave rise to a number of interesting issues and mobilizations: multistakeholderism; civil society coordination; the Communication Rights in the Information Society (CRIS) campaign; NGO partnerships and coordination; and global media governance policy-related initiatives, among other issues. Internet governance has also generated a lot of academic attention in the light of the contested nature of the incumbent body ICANN, the Snowden revelations, and the positions taken by countries that belong to the BRICS consortium who have, at least provisionally and tentatively, begun exploring alternatives to ICANN (see Chapter 5).

Notwithstanding the key importance of both these events, there has been another extremely important initiative related to information rights and access to information, specifically for the visually impaired, that has led to what is known as the Marrakesh Treaty.[4] This relatively poorly

---

[3] Mueller, M.L., B.N. Kuerbis, and C. Page. 2007. 'Democratizing Global Communication?: Global Civil Society and the Campaign for Communication Rights in the Information Society', *International Journal of Communication* 1: 267–96; O'Siochru, S. 2004. 'Will the Real WSIS Please Stand Up?: The Historic Encounter of the "Information Society" and the "Communication Society"', *International Communication Gazette* 66(3–4): 203–24; Raboy, M. 2004. 'The World Summit on the Information Society and Its Legacy for Global Governance', *Gazette* 66(3–4): 225–32; and Thomas, P.N. 2006. 'The Communication Rights in the Information Society (CRIS) Campaign: Applying Social Movement Theories to an Analysis of Global Media Reform', *International Communications Gazette* 68(4): 291–312.

[4] WIPO. 'Marrakesh Treaty to Facilitate Access to Published Works for Persons Who Are Blind, Visually Impaired, or Otherwise Print Disabled, Adopted by the Diplomatic Conference to Conclude a Treaty to Facilitate Access to Published Works by Visually Impaired Persons and Persons with Print Disabilities in Marrakesh, on June 27, 2013', available at http://www.wipo.int/treaties/en/text.jsp?file_id=301016#art3; accessed on 22 May 2018.

cited Treaty in the annals of communication rights is a rare success story, since it does have the potential to facilitate unconditional access to formats and the right to copy material and reproduce it in accessible formats without getting copyright clearance that, until recently, has not been the case. The discussions to facilitate access to published works for blind, visually impaired, and print-disabled people took place under the framework of another UN agency, the WIPO, in 2013. The Treaty was ratified on 27 June 2013 and adopted by 79 member countries of WIPO.

The relative absence of information and writings on this event from a media advocacy perspective remains a gap and is indicative of the following fault lines in global media advocacy. First, global media advocacy tends to be dominated by media NGOs and as was the case at the WSIS, there was very little involvement of other non-media organizations. I have argued elsewhere that global media advocacy requires support from a broad spectrum of organizations related to human rights in order for it to become successful.[5] James Boyle has called for an 'environmentalism for the Net', that is, a movement that is broad-based, globally recognized, and organized and that has the 'muscle' to contribute to the making of policy related to the Internet; and I think that global media advocacy too needs to be broad-based as a movement and connect to and involve other fraternal organizations and movements in order for it become successful on the world stage.[6] An exception to this rule is the broadly based Access to Knowledge (A2K) movement, which began as a loose coalition in 2004 and is involved in a variety of IP-related issues, including access to retroviral drugs and access to knowledge for those with disabilities. It was created by the Consumer Project on Technology and is now known as Knowledge Ecology International.

> The A2K movement, which emerged in 2004 as a broad coalition of interest groups, has found common cause with a broad range of groups working on issues like that mentioned above. Such groups include AIDS activists working on Access to Medicines, computer programmers working on open law

[5] Thomas, 'The Communication Rights in the Information Society (CRIS) Campaign'.

[6] Boyle, J. 1997. 'A Politics of Intellectual Property: An Environmentalism for the Net', *Duke Law Journal* 47(1): 87–116.

coalescing around the notion of Free Culture, librarians promoting access to information, farmers' rights advocates in developing countries protesting seed patents, and others still. This diverse set of transnational activists, scholars, policymakers, and private sector innovators have converged upon a unique identity in a collective critique of propertization and control over information in the prominent industries of the knowledge economy.[7]

Pushing the WIPO to embrace a development agenda, the A2K movement has specifically focused on IP-related issues across a number of sectors, including agriculture, access to life-saving drugs, traditional knowledge, and software. The strength of the movement stems from the fact that it is a broad coalition, and advocates have used this strength to lobby for change in WIPO structures and open space up for access to knowledge initiatives, especially within the global library movement.[8] The fact that WIPO, at its 2010 General Assembly, agreed to adopt the proposed development agenda, monitor its implementation sectorally within WIPO, engage with the 45 recommendations, and support activities worth US$28 million can be considered achievements for the coalitions who have been involved in this movement.[9] There are continuing issues related to NGO involvement in WIPO given the fact that WIPO is yet to formally embrace MSH as a principle; nevertheless, the fact that the World Blind Union (WBU) played an important role in negotiations leading up to the Marrakesh Treaty for the Visually Impaired does suggest that there are spaces and opportunities for negotiation outside of formal multistakeholder approaches. However, as Bannerman has observed: 'The Marrakesh Treaty represents the first time an international copyright treaty has been established with a primary purpose of establishing an international principle of access to knowledge. In that sense, it might be seen as a watershed moment in the history of international copyright

[7] Vetter, T. and E. Katz. 2007. 'Access to Knowledge in the Information Society', Draft Paper, International Institute for Sustainable Development, Manitoba, p. 6, available at http://www.iisd.org/pdf/2007/igsd_access.pdf; accessed on 11 May 2018.

[8] Kapczynski, A. 2008. 'The Access to Knowledge Mobilization and the New Politics of Intellectual Property', The Yale Law Journal 117(5): 804–85.

[9] See Bannerman, S. 2016. International Copyright and Access to Knowledge. Cambridge: Cambridge University Press.

and a true victory for advocates of the visually impaired and the A2K movement.'[10]

Access to knowledge is, however, not a straightforward issue given that the demand by A2K advocates for all knowledge to be in the public domain can be construed as a fundamentalist position from the point of view of traditions, in which access to knowledge is invested with spirituality and continuities, and is culturally regulated. Allison Fish, commenting on her work in India among religious practitioners of yoga, learned that the issue was not really about whether yoga should or should not be in the public domain but to take it seriously and acknowledge the fact 'that the central goal of yoga is the liberation of the individual self and that a practitioner approaches this goal through a commitment to a daily practice of living explicitly structured through his or her hierarchical relationship with an expert guru who acts as a gatekeeper over spiritually powerful knowledge.'[11]

The second problem has been that the WSIS was an all-consuming event that, unfortunately, did not result in an impetus for change. Arguably, it sapped the energies of cash-strapped media NGOs. For most media NGOs, international advocacy remains a luxury that they can ill afford given other pressing deadlines related to projects. Third, the WSIS also led to focused and exclusive academic attention, arguably to the detriment of other critical, on-going negotiations, such as the treaty for the blind, the convention on cultural diversity, the A2K and UNESCO-related meetings on gender and media.

Background to the Marrakesh Treaty

The WHO statistics indicate that there are 285 million people whose vision is impaired. This includes 39 million who are blind and 246 million who have low vision. Ninety per cent of the visually impaired live in the developing world.[12] People with visual disabilities have been

---

[10] Bannerman, *International Copyright and Access to Knowledge*, p. 182.

[11] Fish, A. 2014. 'The Place of "Culture" in the Access to Knowledge Movement', *Anthropology Today* 30(5): 8.

[12] WHO, 'Visual Impairment and Blindness', 11 October 2018, available at http://www.who.int/mediacentre/factsheets/fs282/en/.

among communities who have suffered from a 'book famine', resulting in a lack of access to knowledge that is critical to social and cultural well-being. This famine has been caused due to lack of availability of books in formats such as Braille, large print, and audio editions, as well as copyright restrictions that have been placed on international exchange of material in these formats and the right of the visually impaired to copy books of their choice without taking permission from the rights owners. Only five per cent of the world's books are available in accessible formats and it is estimated that in the developing world, this falls to less than one per cent.[13] The UN Convention on the Rights of Persons with Disabilities, 2006, outlines its commitment to accessibility in Article 9, 'Freedom of Expression and Opinion, and Access to Information', as follows:

a) Providing information intended for the general public to persons with disabilities in accessible formats and technologies appropriate to different kinds of disabilities in a timely manner and without additional cost;

b) Accepting and facilitating the use of sign languages, Braille, augmentative and alternative communication, and all other accessible means, modes and formats of communication of their choice by persons with disabilities in official interactions;

c) Urging private entities that provide services to the general public, including through the Internet, to provide information and services in accessible and usable formats for persons with disabilities;

d) Encouraging the mass media, including providers of information through the Internet, to make their services accessible to persons with disabilities;

e) Recognizing and promoting the use of sign languages.[14]

However, despite the stated commitments, it did not have the means to make legally available these formats until the Marrakesh Treaty.

The key device in the IP armoury that is used to restrict access has been copyright. The issue of maintaining a balance between the rights of creators and public interests has been controversial to say the least, given

[13] 'Increased Access to Alterative Formats'. 2018. *VisionAustralia*, 6 July, available at https://www.visionaustralia.org/community/news/06-07-2018/Copyright-Act-changes; accessed on 26 March 2019.

[14] 'UN Convention on the Rights of Persons with Disabilities. 2006. Article 21: Freedom of Expression and Opinion, and Access to Information', available at http://www.un.org/disabilities/default.asp?id=281; accessed on 11 May 2018.

the many contemporary moves to lengthen the terms for copyright at the expense of the public. One of the contentious and yet to be resolved issues in the area of copyright, for instance, is the issue of 'authorship', a concept that was developed from Enlightenment scholars such as John Locke in his treatise on the 'natural rights of man' and given legal status through the Copyright Act of April 1710, enacted in the UK, that gave protection to individual authors against print piracy. Today, 'authorship' does not accrue either to the individual or to the collective authors of traditional texts, software, and other cultural products, but to their corporate employers. The migration of rights originally invested in individuals to corporations and the attempts to legalize such translations have been primary strategic means employed by corporations in their bids to take control over the knowledge economy. The Berne Convention that remains the standard basis for the legality of copyright includes what is commonly referred to as the three-step test that needs to be taken to make the case for any 'exceptions' related to copyright. These steps have been incorporated into the TRIPS Agreement, the WIPO Copyright Treaty, and national legislations.[15] The three steps are as follows: (i) 'certain special cases'; (ii) which do 'not conflict with a normal exploitation' of the copyright material; and (iii) do 'not unreasonably prejudice the legitimate interests' of the rights holder. However, despite the stated exceptions, it has been difficult to make a case for such exceptions because of the pressure from rights holders as well as from governments. Nwankwo highlights some of their concerns:

> As the debate continues over expanding exceptions and limitations for the benefit of VIPs, copyright owners have voiced their concerns in opposition to the proposal. Both in the US and the EU, these copyright owners, including authors and publishers, have maintained that the proposal is prejudicial to the existing international copyright framework. They have argued that such an exception will open the flood-gate for people who are not visually impaired to pirate their works. They equally insist that where resources are already scarce, the existence of copyright exemptions further reduces incentives to

[15] Knights, R. 2001. 'Limitations and Exceptions Under the "Three-Step-Test" and in National Legislations—Differences between the Analogue and Digital Environments', WIPO, available at http://www.wipo.int/edocs/mdocs/copyright/en/wipo_cr_mow_01/wipo_cr_mow_01_2.pdf; accessed on 2 May 2018.

invest in the production and distribution of works in accessible formats into the market.[16]

The Politics of Digital Rights Management

One of the consequences of the current interpretation is the negation of the social, public value of IP, a value that exceeded its value as private property. 'Authorship' continues to be a problematic concept, particularly so in the context of new, digital technologies of reproduction. Digital technologies have unfixed the singular location, fixed materiality, and objectivity associated with products related to the previous generation of technologies. The appropriation and transformation of digitally manipulated material, access to 'sampling' technologies, intertextual mixing and matching, and net-based peer-to-peer sharing of popular culture have further complicated the notion of authorship. However, industry has responded to what it sees as a threat to its profit margins by incorporating Digital Rights Management (DRM) standards in hardware and software. The DRM places restrictions as well as limits and controls on what users can do with hardware and software; in other words, it manages user access. Apple's iBook and Amazon's Kindle are both DRM-enabled, thus restricting access to and adaptability for those with visual impairments. The fact that the US Digital Millennium Copyright Act (DMCA) has criminalized encryption circumvention devices—for example, the Advanced eBook Processor (AEBPR) that enables people with disability to copy Adobe eBooks and other formats into more user-friendly formats—has made access all the more difficult for people with disabilities. The AEBPR figures out and removes user restrictions on DRM-enabled software. This was the reason why Dmitry Sklyarov, employed with Moscow-based ElcomSoft and one of the creators of AEBPR, was arrested. Sklyarov was the first victim of the DMCA. As Parloff, who is critical of ElcomSoft's marketing and circumvention practices, explains:

[16] Nwankwo, I.S. 2011. 'Proposed WIPO Treaty for Improved Access for Blind, Visually Impaired, and Other Reading Disabled Persons and Its Compatibility with TRIPS Three-Step Test and EU Copyright Law', *Journal of Intellectual Property, Information Technology and Electronic Commerce Law* 2(3): 211.

Sklyarov simply made software that removes certain security protections from the Adobe eBook Reader, enabling people to engage in numberless, marvellous, invaluable, noninfringing uses. 'A blind person,' for instance, could use it to activate Adobe's 'read-aloud function' in order 'to listen to a book,' even if Adobe had, at a publisher's instructions, disabled that function for a particular title. Alas, has helping the blind to read become a crime in our country? Sklyarov's wonderful creation also enables people to make back-up copies of e-books, explains the Electronic Frontier Foundation's Web site. It also enables them to transfer an e-book from an old computer to a new one. Best of all, Sklyarov's software only works on lawfully purchased e-books. So what in the world could the prosecutors and Adobe and the Association of American Publishers—which has applauded the prosecution—possibly be so upset about?[17]

One of the issues with DMCA's stance on exemptions is the fact that it is very rarely that they concede the need for any change. The Copyright Office in the USA convenes a rule-making event every three years in order to update the DMCA and four rule-making conventions have been organized in 2003, 2006, 2009, and 2012. At the latest rule-making event, exemptions were narrowed down even further, including the use of eBooks by the visually impaired who are required to ensure that 'the rights owner is remunerated, as appropriate, for the price of the mainstream copy of the work as made available to the general public through customary channels' when electronically available literary works are adapted so as to make available its read-aloud functionality.[18] This narrowing down goes against the spirit of the Marrakesh Treaty and highlights a major issue today: the jurisdiction and relative 'power' of legislations and treaties such as the Marrakesh Treaty on the one hand, and the DMCA on the other. These restrictions and others were challenged in the USA by the American Council for the Blind and the American Foundation for the Blind,[19] particularly the requirement that in the light of device and format

---

[17] Parloff, R. 2001. 'Free Dmitry? Spare Me: Why the FBI was Right to Arrest the Internet's Latest Martyr', 1 August, p. 2, available at https://www.law.upenn.edu/law619/f2001/week09/parloff_dmca.pdf; accessed on 13 May 2018.

[18] See Gray, M. 2014. 'New Rules for a New Decade: Improving the Copyright Office's Anti-circumvention Rulemakings', *Berkeley Technology Law Journal* 29(4): 793.

[19] Richert, M. 2014. 'Before the United States House of Representatives Judiciary Committee Subcommittee on Courts, Intellectual Property, and

incompatibilities, the visually impaired had no choice but to purchase these devices in order to access knowledge. In other words, DRM restrictions have continued to make life difficult for the visually impaired. Interestingly enough, in response to the opposition by the Joint Creators and Copyright Owners, the NTIA took the side of the visually impaired:

> Requiring visually impaired Americans to invest hundreds of dollars in an additional device (or even multiple additional devices), particularly when an already owned device is technically capable of rendering literary works accessible, is not a reasonable alternative to circumvention and demonstrates an adverse effect of the various access controls used. Therefore, NTIA supports adoption of the expanded exemption for the next three years, and strongly encourages the market to obviate any future need for this exemption by making literary works more accessible to users with disabilities, ideally in an interoperable manner.[20]

The power of industry that lobbies to protect the core features of the DMCA, in particular its anti-circumvention stance, is bound to remain a difficult issue in IP negotiations pursued by the USA at global forums, such as the WIPO Development Agenda. Among the 45 recommendations adopted by this Agenda is: '17. In its activities, including norm-setting, WIPO should take into account the flexibilities in international intellectual property agreements, especially those which are of interest to developing countries and LDCs.'[21] Whether or not the hard-won deals at Marrakesh will be compromised remains an issue in the context of negotiations related to the Trans-Pacific Partnership, a trade agreement between 12 Pacific nation states, in which the USA has stood by strict

---

the Internet Hearing on Chapter 12 of Title 17: Testimony of Mark Richert, Director of Public Policy, American Foundation for the Blind', 17 September 2014, available at http://judiciary.house.gov/_cache/files/c1d41f70-cb2b-4529-a85a-1e5620468971/09.17.14-testimony-afb.pdf; accessed on 14 May 2018.

[20] Cox, K. 2013. 'Marrakesh Note 4: The 2012 U.S. Copyright Office Decision Regarding Technological Protection Measures, Including Discussion of Commercial Availability of Accessible Works', 7 June 2013, p. 4, available at http://keionline.org/sites/default/files/TPMs_CommercialAvailability.pdf; accessed on 15 May 2018.

[21] WIPO, 'The 45 Adopted Recommendations under the WIPO Development Agenda', available at http://www.wipo.int/ip-development/en/agenda/recommendations.html#b; accessed on 30 May 2018.

rules related to the circumvention of technological protection measures (TPMs) and has not included a clause on exemptions. This inconsistency with the Marrakesh Treaty does indicate that the default position for USA negotiators, irrespective of limited rule-making exemptions, will be iron-clad anti-circumvention restrictions.

The WIPO Treaty

When the WBU drafted a submission that was presented by the governments of Brazil, Ecuador, and Paraguay at WIPO on the need for a treaty related to access for the visually impaired, the immediate reaction from a range of apex bodies related to the creative and copyright industries in the USA was negative. The Association of American Publishers, the Motion Picture Association of America, the Recording Industry Association of America, the Software and Information Industry Association, along with other organizations, made submissions to the US Copyright Office in which they began with preambles that expressed support for access and WIPO, but also included statements that questioned the wisdom of the initiators of this treaty including WIPO. Undoubtedly, the publishing industry in the USA has played a key role in influencing the US position given that in 2010 alone, they 'generated a net revenue of $27.9 billion'.[22] They have consistently argued that any dilution of DRM would result in piracy, to an opening of the floodgates and losses to the copyright industries. The following excerpt is from the submission made by the Software and Information Industry Association in 2009:

> The bilateral approach of mutual cooperation working within the marketplace is the best way to develop the technological solutions to the specific issues related to facilitating access to copyrighted works for the blind and visually impaired. ... It would be premature and counter-productive to prescribe in treaty form the very technologies and market that is facilitating, for the first time in human history, the very accessibility long sought by blind and visually impaired individuals. There is a real danger that these cooperative efforts

---

[22] Bertlesman, M. 2012. 'The Fight for Accessible Formats: Technology as a Catalyst for a World Effort to Improve Accessibility Domestically', *Syracuse Journal of Science & Technology Law* 27(3): 52.

could be adversely affected if WIPO or any government were to step in and attempt to create and implement copyright-specific mandates or exceptions that bind the hands of the stakeholders, especially those being proposed in the draft Treaty.[23]

Their attitude and position remained unchanged for much of the negotiations leading up to the Treaty and beyond. It is, therefore, even more remarkable that the Treaty was established given the amount of opposition to it. One of the interesting complexities of this Treaty is the position of WIPO itself that represents the interests of its members along with the interests of industry. In a 2006 submission to the Standing Committee on Copyright and Related Rights, WIPO, Nic Garnett unambiguously stated that 'at present, neither the market nor technology appears to be supporting a basis for facilitating the access to information by visually impaired people in a way that is consistent with the general standards for the full social and economic integration of people with disabilities.'[24] Interestingly enough, a presentation in Geneva, this time in 2008, by the very same person, made no mention of the role of the 'market' in obstructing access for the visually impaired to digital texts.[25] Following the WBU's call for a Treaty for Visually Impaired People in 2006, there were at least four instances where member countries tabled proposals for treaties at WIPO between 2009 and 2010, including

---

[23] 'Submission of The Software & Information Industry Association in Response to the Notice of the U.S. Copyright Office and the U.S. Patent and Trademark Office on October 13, 2009 Requesting Comments on the Topic of Facilitating Access to Copyrighted Works for the Blind or Persons with Other Disabilities', 13 November 2009, pp. 1–4, available at http://siia.net/index.php?option=com_docman&task=doc_view&gid=2331&Itemid=318; accessed on 11 May 2018.

[24] Garnett, N. 2006. 'Automated Rights Management Systems and Copyright Limitations and Exemptions', WIPO Standing Committee on Copyright and Related Rights, Geneva, 1–6 May, p. 33, available at http://www.wipo.int/edocs/mdocs/copyright/en/sccr_14/sccr_14_5.pdf; accessed on 9 May 2018.

[25] Garnett, N. 2008. 'Informative Session on Limitations and Exemptions, Geneva November 3, 2008, Presentation of the Study Entitled "Automated Rights Management Systems and Copyright Limitations and Exemptions"', available at http://www.wipo.int/edocs/mdocs/copyright/en/sccr_17/sccr_17_www_111452.pdf; accessed on 13 May 2018.

a joint proposal made by the WBU and Brazil, Paraguay, and Chile, a proposal submitted by nations from Africa, a proposal from the EU, and another from the USA. The baseline proposal for the Treaty which set the standard for ensuing discussions was the joint proposal. While the first two proposals supported the spirit of WIPO's Development Agenda, the proposals from the EU and the USA were more measured. The EU proposal made the case for a 'trusted intermediary' that is acceptable to both right holders and to user communities to negotiate, while the US proposal restricted the availability of accessible formats and did not allude to the WIPO Development Agenda. The notion of the 'trusted intermediary' had its origins in the Chafee Amendment to the Copyright Act, 1976, USA, that remains the most progressive legal means to expand access to printed works for both the visually impaired and the disabled in which the role of the 'authorized entity', an organization dedicated to access services, is key to negotiations between the stakeholders. However, the Chaffee Amendment too limited the 'specialized formats' in which visually impaired persons (VIPs) could access texts.[26] In the USA, till date, fair use and the 'Chafee Exemption' remain the two limitations to any author's exclusive rights.

In 2011, a joint submission was made to WIPO that reflected a negotiated settlement. Communication rights scholars will be familiar with a previous example of a negotiated settlement, the original call for a New World Information Order (NWIO) as a flip side of the New World Economic Order (NWEO), displaced by the New World Information and Communication Order (NWICO) supported by UNESCO, which ignored critical issues inclusive of the 'decolonisation of information' and the need for a new economic order. Sean Williams, in an article that explores the movement towards a treaty, highlights the advantages to the developed countries in the 2011 joint proposal:

> many of the more restrained features pushed for by the developed countries appear to have prevailed in the latest VIP treaty incarnation. For example, the 2011 proposal is framed as non-binding rather than binding, and it sets forth a rather narrow definition of VIPs. Moreover, it retains reference

[26] See Fitzpatrick, S. 2014. 'Setting Its Sights on the Marrakesh Treaty: The U.S. Role in Alleviating the Book Famine for Persons with Print Disabilities', *Boston College International and Comparative Law Review* 37(1): 139–72.

to the developed countries' use of trusted intermediaries (now referred to as 'authorized entities'). Additionally, the 2011 proposal explicitly refuses to bar contracting around the treaty as the developing countries originally sought in their initial proposals, and the proposal allows countries to limit VIP exceptions 'to published works which, in the applicable special format, cannot be otherwise obtained within a reasonable time and at a reasonable price.' Lastly, the 2011 proposal allows countries to require remuneration for accessible works distributed under the treaty.[27]

Notwithstanding these drawbacks, the 2013 Treaty importantly provides the necessary flexibility for national copyright law to create the space for exemptions. The Berne Three-Step-Test remains the international standard for assessing exceptions to copyright. It has been incorporated into a number of international treaties including the Agreement on the Trade Related Aspects of Intellectual Property Rights (TRIPS), 1994, and the WIPO Copyright Treaty and WIPO Performances and Phonograms Treaty, 1996.[28] While copyright legislations in some developed countries do include exemptions, most legislations in the developing world, where most visually impaired people live, do not have such flexibilities. There are four important achievements of the Treaty. First, the Treaty enables VIPs and the organizations that they relate to, to make copies of works without getting the permission of rights holders. This is an important advancement given that the politics of permission often obstructed any meaningful negotiations. Organizations seeking permission were often given impossible restrictions either in what formats they could use or the high compensation they were asked to pay the rights holders. Second, it allows for the import and export of accessible versions of printed works regardless of borders. This, again, is an important advancement given that there were strict restrictions on the exchange of works for VIPs, especially if works were based on the circumvention of DRM. An interesting example of an organization that has created digital libraries for VIPs, such as Bookshare.org, is the Digital

[27] Williams, S. 2012. 'Closing in on the Light at WIPO: Movement towards a Copyright Treaty for Visually Impaired Persons and Intellectual Property Movements', *University of Pennsylvania Journal of International Law* 33(4): 1061.

[28] See '4. The Case for Fair Use: Fair Use Complies with the Three Step Test'. n.d. Australian Government: Australian Law Reform Commission, available at https://www.alrc.gov.au/publications/4-case-fair-use/fair-use-complies; accessed on 26 March 2019.

Accessible Information System (DAISY) Consortium consisting of the world's leading libraries for the disabled. Their objective is to create texts for the VIPs in the DAISY open format based on the DAISY standard. DAISY is a flexible format that allows for navigation between content—text, audio, image—and its structure. More often than not, a DAISY audio text is available in an MP3 format and, typically, a user can listen to the table of contents and link directly to the chapter of his/her choice. There is also the DAISY full-text book, which the VIPs can read using a voice synthesizer or Braille display. As it is explained on their website:

A DAISY book can be explained as a set of digital files that includes:

- One or more digital audio files containing a human narration of part or all of the source text;
- A marked-up file containing some or all of the text (strictly speaking, this marked-up text file is optional);
- A synchronization file to relate markings in the text file with time points in the audio file; and
- A navigation control file which enables the user to move smoothly between files while synchronization between text and audio is maintained.

The DAISY Standard allows the producing agency full flexibility regarding the mix of text and audio ranging from audio-only, to full text and audio, to text-only.[29]

The treaty makes it easier for the DAISY standard to be adapted around the world and for an exchange of material.

Third, the treaty extends and gives meaning to a range of understandings of the 'beneficiary' of the treaty. It recognizes the fact that there are different gradations of visual impairment, contexts, and conditions that give rise to such impairments. As Article 3 states:

A beneficiary person is a person who:

(a) is blind;
(b) has a visual impairment or a perceptual or reading disability which cannot be improved to give visual function substantially equivalent to that of a

---

[29] DAISY Consortium. 'Creating the Best Way to Read and Publish', available at http://www.daisy.org/about_us; accessed on 22 May 2018.

person who has no such impairment or disability and so is unable to read printed works to substantially the same degree as a person without an impairment or disability; or

(c) is otherwise unable, through physical disability, to hold or manipulate a book or to focus or move the eyes to the extent that would be normally acceptable for reading; regardless of any other disabilities.[30]

This broadening of understanding of a VIP has enabled flexibilities and made it possible for a range of beneficiaries to be included. Fourth, the treaty provides the legal right to circumvent TPMs, such as DRM. In the context of the digital environments that we live in and the fact that major original and derivative works are available in digital formats, this is an important advancement, though it is bound to be contested at different forums by the copyright industries. The circumvention of DRM will enable VIPs to access reliable screen readers and text-to-audio files. The next step in the process is for the Treaty to be ratified by countries and its recommendations incorporated into amended national copyright legislations. While a number of countries (79) had signed the Treaty by July 2014, it could only come into effect when a minimum of 20 countries ratified it. The GoI was the first to ratify the Treaty in June 2014, followed by El Salvador in October. The European Commission had proposed its ratification in a press release dated 21 October 2014.[31] This is an important beginning and organizations throughout the world are pressuring their national governments to ratify the treaty.

Access for the Visually Impaired in India

With around 62.6 million people with visual impairments, VIPs in India have a lot to gain from India's ratification of the WIPO Treaty.[32] Indian

---

[30] WIPO. 'Marrakesh Treaty to Facilitate Access to Published Works for Persons Who Are Blind, Visually Impaired, or Otherwise Print Disabled'.

[31] 'European Commission Proposes Ratification of Marrakesh Treaty to Facilitate Access to Books for Visually Impaired Persons', European Commission, Brussels, 21 October 2014, available at http://europa.eu/rapid/press-release_ IP-14-1185_en.htm?locale=en; accessed on 11 May 2018.

[32] WHO. 2010. 'Global Data on Visual Impairments', Geneva, available at https://www.who.int/blindness/GLOBALDATAFINALforweb.pdf; accessed on 26 March 2019.

NGOs, such as Inclusive Planet, the CIS, National Association for the Blind (Delhi), Saksham, National Federation for the Blind, Samarthanam (Bengaluru), Mitrajyoti (Bengaluru), the Alternative Law Forum, and others, have played an important role in both domestic and international advocacy. Certain state governments, such as that of Kerala, have also played an important role in creating accessible software and hardware for people with disabilities. The nationwide Right to Read campaign, launched in 2007, has helped to increase awareness in India among the lay public, and among authors and publishing houses, on the need for VIP access. Also, the issue of accessibility has played a key role in pressuring the GoI to take a pro-access stance.

The partnership between the state and civil society is best illustrated by the relationship between the government of Kerala's IT@School project and the NGO, the Society for Promotion of Alternative Computing and Employment (SPACE). SPACE has played an important role in developing FOSS-based software—the IT@School GNU/LINUX Lite operation system that can be used on computers with low memory, FOSS-based training programmes for teachers, resource centres, and FOSS support groups.[33]

With bipartisan political support from the Communist and Congress governments, Kerala is now home to numerous FOSS-based public-sector software initiatives, including IT@School, Insight, and the Akshaya tele centres; 600+ open source learning centres across the state;[34] Health Information Systems;[35] the C-DAC; the National Informatics Centre; the Centre for Advanced Training in Free and Open Source Software; the International Centre for Free and Open Source Software that was established in 2011; and SPACE. The intent is to democratize software solutions, bridge the digital divide, and empower local communities via

---

[33] See 'The Story of IT@School's Migration to Free Software'. n.d., available at   https://www.space-kerala.org/files/Story%20of%20SPACE%20and%20 IT%40School.pdf; accessed on 26 March 2019.

[34] Krishnaswamy, G. and D. Marinova. 2011. 'FOSS in Education: IT@ School Project, Kerala, India', available at https://pdfs.semanticscholar.org/5c66/ f4703de485f6d9bc877090c5f37909da4fba.pdf; accessed on 26 March 2019.

[35] Puri, S.K., S. Sahay, and J. Lewis. 2009. 'Building Participatory HIS Networks: A Case Study from Kerala, India', *Information and Organisation* 19(2): 63–83.

digital literacy initiatives. Kerala celebrated 10 years of state-sponsored FOSS initiatives in 2011, although the history of FOSS in this state is at least a decade older.[36]

Disability remains a major social and cultural issue and people with disabilities in India are among the most culturally and socially marginalized people in the country. While Constitutional guarantees and special legislations do exist—such as the Persons with Disabilities (Equal Opportunities, Protection and Full Participation) Act, 1996, including Sections 27 and 28 that require the government to both design and invest in research in assistive devices and make accessible resources in relevant formats; and the Rights of Persons with Disabilities Act, 2016, that supports employment-related protections for people with disabilities—there has been a disconnect between policy and practice. However, and over the last two decades, NGOs have increasingly played an important role in the provisioning of disability services and there are close to 3,000 registered NGOs working on disability-related issues.[37] There are a number of disability activists in India including Zamir Dhale of Sense International and the efforts made by the lawyer-activist Rahul Cherian, whose untimely demise in February 2013 has been a loss to both international and national advocacy related to rights for people with disabilities. Cherian's advocacy efforts at WIPO have been documented and are available on the CIS website.[38]

As a result of nation-wide advocacy by disability activists and NGOs in India, the Copyright Amendment Bill's, 2012, access commitments for the visually impaired are among the most progressive in the world today. The amendments made to the 2012 Bill have turned access for the visually impaired into a national, human right.[39] As Prakash has observed:

---

[36] Thomas, P.N. 2014. 'Public Sector Software and the Revolution: Digital Literacy in Communist Kerala', *Media, Culture & Society* 36(2): 258–68.

[37] See Divyangjan. 2014. 'List of Registered NGO, Department of Empowerment of Persons with Disabilities', available at http://disabilityaffairs. gov.in/content/page/list-of-registered-ngo.php; accessed on 26 March 2019.

[38] See http://cis-india.org/@@search?SearchableText=Rahul+Cherian; accessed on 9 May 2018.

[39] Ministry of Social Justice and Empowerment. 2012. 'The Draft Rights of People with Disabilities Bill, 2012'. Department of Disability Affairs, Ministry of Social Justice and Empowerment, GoI, pp. 1–83, available at http://socialjustice. nic.in/pdf/draftpwd12.pdf; accessed on 10 February 2013.

Section 52(1)(zb) allows any person to facilitate access by persons with disabilities to copyrighted works without any payment of compensation to the copyright holder, and any organization working [for] the benefit of persons with disabilities to do so as long as it is done on a non-profit basis and with reasonable steps being taken to prevent entry of reproductions of the copyrighted work into the mainstream.[40]

This Bill foreshadowed the WIPO Treaty for the Visually Impaired.

Moreover, standard printed books can now be converted into accessible formats, including Braille, synthetic audio, text readable by screen reader, and so on, without having to take prior permission from the copyright holder. The fact that the requirement for prior permission has been removed is a major victory for VIPs in India.[41] Clause (zb) allows for fair use, including the adaptation, reproduction, and production of material for educational, personal use, and the sharing of such material among people with disabilities. This can be carried out by any organization involved either primarily or secondarily in disability rights. This clause nullifies the previous requirement that any organization applying for a compulsory licence to reproduce material for the disabled had to be registered, work solely in this area, and be recognized by the government. Furthermore, people with disabilities cannot, by themselves, apply for such licences.[42] Section 31(B) allows any person working in the area of disability to apply for a compulsory licence to publish any copyrighted work for the benefit of the disabled. Furthermore, there is now a correspondence between the amended copyright act and domestic disability legislations, such as the Draft Rights of People with Disabilities Bill, 2012.

51. Access to Information and Communication Technology

(1) Appropriate governments and establishments shall take measures to ensure that:

---

[40] Prakash, P. 2012. 'Analysis of the Copyright (Amendment) Bill, 2012', CIS, 23 May, available at http://cis-india.org/a2k/blog/analysis-copyright-amendment-bill-2012/; accessed on 10 February 2013.

[41] Pillai, P.R. 2012. 'Accessible Copies of Copyright Work for Visually Impaired Persons in India', *Creative Education* 3(26): 1060–2.

[42] Saikia, N. 2010. 'Disability and the Indian Copyright (Amendment) Act, 2010', *Social Science Research Network*, p. 8, available at http://papers.ssrn.com/sol3/papers.cfm?abstract_id=1600621; accessed on 10 February 2013.

a. All content in whichever medium whether audio, print or electronic shall be made available to persons with disabilities in accessible format;

b. Persons with disabilities have access to electronic media by providing for audio description, sign language interpretation and close captioning;

c. Accessibility to telecommunication services where telecommunications will include any kind of transmission of information of the user's choosing without change in form or content of information as sent or received;

d. Electronic goods and equipment of everyday use shall follow the principles of universal design;

e. Schemes are formulated or amended to ensure affordable access to Information and Communication Technology & Electronics for persons with disabilities in rural as well as urban areas;

f. Incentives and concessions are provided to support existing websites to make them accessible to persons with disabilities.[43]

Arguably, both the GoI's support for an amendment of its copyright law in favour of copyright-free access to material by the visually impaired and its support for the WIPO Treaty are examples of what Polanyi has described as a 'double movement'. In his book, *The Great Transformation*, the double movement has been described as the means by which a re-embedding of the 'social' was accomplished as a response to the unnatural fiction and reality of the primacy of the unregulated market.[44] The welfare state's involvement in redistribution and its exertion of state control over labour and finance markets did restrict the potential effects of some of the worst excesses of predatory forms of capitalism and the market economy. In the specific case of access for the visually impaired, it reflects a clear counter-movement against the commodification of IP and the extension of copyright in perpetuity over works reflected in global and national IP legislations. Given the moral issue at stake, India being home to the largest population of visually impaired in the world, and

[43] Government of India. 2012. 'The Draft Right of Persons with Disabilities Bill', Ministry of Social Justice and Empowerment, Department of Disability Affairs, p. 37, available at http://www.dnis.org/draftpwd12.pdf; accessed on 26 March 2019; and Thomas, P.N. 2014. *Copyright and Copyleft in India: Between Global Agendas and Local Interests*, in Matthew David and Debora Halbert (eds), *The Sage Handbook of Intellectual Property*. London, Thousand Oaks, and New Delhi: SAGE, pp. 355–69.

[44] Polyani, K. 1944. *The Great Transformation: The Political and Economic Origins of Our Time*. New York: Farrar & Reinhardt.

pressure from strong lobby groups within India, the Indian government, through its legislative actions, has arguably brought some balance to the very idea of copyright as a limited right and access as a human right, especially for those whose quality of life has been impaired by a lack of access to the world's knowledge. What is clear from this example is that the GoI's actions were motivated by the need to bring back a semblance of balance to copyrights for a very specific community who had traditionally been restricted from accessing knowledge. Arguably, the net result of State actions has been to find solutions outside of the market as it were. Perhaps it would be more accurate to state that in this instance, the market imperative has been suppressed in favour of the collective social. Against the overtly pro-MNC IP leanings of the current government, these pro-people acts may seem inconsequential given the relative marginality of the group in question. However, such incremental wins need to be seen in the context of the double movement manifesting itself in the everyday lives of people through access to employment, food, information, and so on, within the welfare economy.

***

The WIPO Treaty is the beginning for new possibilities of access to and the modifiability of digital works by VIPs. Despite the need for more countries to ratify the treaty, it represents a significant move forward; just as pharmaceutical companies were shamed into relenting on the need for generic drugs for patients with acquired immune deficiency syndrome (AIDS) during the Doha round of WTO negotiations in 2003, there are indications that some copyright industries have relented and have become open to negotiate the 'moral' rights of IP. The launch of the project, Your Voice, Their World, the largest digital access library project for VIPs in India by the Japanese technology firm, Omron, and the National Association for the Blind, New Delhi, along with the Accessible Books Consortium (ABC) at Marrakesh by WIPO, a multistakeholder project that includes the major publishing associations, the DAISY Consortium, and the WBU, among many other groups, are indicative of the shifts towards the creation of global access for VIPs. Shillipi Singh, in an article, highlights 10 digital and Braille-based inclusion projects for the visually impaired in India, including the Central Library of Audio Books in Indian Languages; Hear2Read, a text-to-speech software that enables reading without seeing; a newspaper in Braille in Marathi language, *Sparshdnyan*, meaning 'knowledge by touch'; and *White Print*, a '64-page monthly magazine [that] is printed at the National Association for the Blind, Mumbai, and circulated across India. The magazine has articles on sports, politics, culture, fashion, technology, inspiring stories of the common man, short stories and even reader contributions', among other initiatives.[45] While formal initiatives aimed at digital inclusion are important, it is clear that incremental progress on access has led to the visually impaired taking matters into their own hands with respect to digital access. In a study on access to and use of social media by the visually impaired in the western Indian state of Rajasthan, Vashishta et al. have observed that piracy is rife:

> It is worth mentioning that many low-income blind participants either did not know about piracy or they did not care about digital rights infringement. A majority of the participants in the formative study did not know about the cost of screen reader software. For them, both NVDA [Non Visual Desk Top Access] and JAWS [Jobs Access With Speech, software for the use of the visually impaired] were freely accessible. Only one user bought a screen reader software for 50 USD from a non-profit organization. Many recent adopters planned to either download it or ask [for] a copy from friends and instructors. The pervasive piracy of screen reader software in India is also reported by other researchers. Even in our voice-based social media platform, we received many messages where participants recorded songs from other playback devices without caring about copyright issues.[46]

The combination of both formal and informal initiatives is bound to contribute to greater access to communication for the visually impaired.

[45] Singh, S. 2017. 'They Can't See, But They Can "Read"', *The Free Press Journal*, 23 April 2017, available at http://www.freepressjournal.in/featured-blog/they-cant-see-but-they-can-read/1056282; accessed on 2 May 2018.

[46] Vashishta, A., E. Cutrell, N. Dell, and R. Anderson. 2015. 'Social Media Platforms for Low-Income Blind People in India', ASSETS '15, Lisbon, Portugal, 26–8 October, p. 12, available at https://www.microsoft.com/en-us/research/wp-content/uploads/2016/02/Vashistha-ASSETS2015-social-media-for-blind.pdf; accessed on 4 May 2018.

Conclusion

Digital India and the Politics and Geopolitics of Information

The chapters in this book have underlined some of the ongoing external and internal contestations that relate to the politics of the digital economy in India. They have attempted to deal with different aspects of the Indian State's acts of omission and commission in its governance of the digital economy. In the context of a democracy that is flawed and far from perfect, the case studies presented here have served to demonstrate that the spread of global digital capitalism is uneven, complex, and contested. The digital exists in many forms and shapes in India and its 'formal' avatars are complemented by a variety of informal digital economies. These informal digital spaces have a life of their own and are integrated into the local, making it difficult to police the official programme of digitization. I have argued that the various responses of the Indian state are symptomatic of a 'double movement'— with neo-liberal policies accompanied by information access initiatives that offer an antidote to the relentless accumulation by dispossession that is a nationwide reality today. However, the role of the State in this context remains contradictory. While State surveillance is tightly focused on the 'war against terrorism', it is only fitfully directed towards the monitoring of IP and other 'subversions' of the knowledge economy. The IP rules in particular are, for various reasons, difficult to implement despite substantive investments by international and domestic trade lobbies and companies linked to the information and cultural industries.

The need for stronger IP protection for the US cultural industries in India continues to be an aspect of the geopolitics of information. Section 301 has been used by the USTR against India ostensibly in retaliation against India's inability to curb a variety of cultural and other piracies. Also, India has consistently been on the USTR's 'priority watch list' between 1989 and 2018. However, curbing cultural piracy is just one aspect of the US pressure to liberalize the Indian economy, which is manifested in attempts to lower perceived high tariffs in industrial imports, the opening up of the banking and insurance sector to foreign investment, and reforms in the fiscal, industrial, and trade sectors. India continued to remain on the Priority Watch List category in 2018 and was also chosen for an 'Out of Cycle Review' in 2014.[1] While the National IP Rights Policy, 2016, was framed to be US-friendly, at the Section 301 review in 2017, major MNCs in the USA and apex bodies such as the US-based International Intellectual Property Alliance cited the lack of concrete policy measures as a reason to keep India within the Priority Watch category. Nonetheless, for the most part, in the post-Bush era, Obama's term, and now Trump's administration, there has been an intention to forge closer ties between India and the USA, not only in the matter of trade but also in terms of security, the fight against terrorism, and the commitment to democracy and rule of law. The deterioration of the USA–Pakistan relationship has occurred in parallel with India's warming relationship with the USA, reflecting a mutual concern against China's economic and territorial ambitions in South Asia. Russia, the USA, and Israel are among the key suppliers of arms to India. Between 2013 and 2017, the largest suppliers were Russia (62 per cent), the USA (15 per cent), and Israel (11 per cent).[2] India and the USA have also come together in joint military exercises directed towards curbing the growing influence of China in the Indian Ocean region. Increasingly, the Indo-Pacific region has become the stage for a shared geopolitical strategy between the USA and India. From the perspective of the USA, India can play a significant role in contributing to and strengthening a joint security architecture aimed at curbing the

---

[1] USTR. 2014. 'USTR Begins Special 301 Out-of-Cycle Review of India', available at https://ustr.gov/about-us/policy-offices/press-office/press-releases/ 2014/October/USTR-Begins-Special-301-Report-Out-of-Cycle-Review-of-India; accessed on 22 August 2018.

[2] See Pandit, R. 2018. 'With 12% of Global Imports, India Tops List of Arms Buyers', *The Times of India*, 13 March, available at https://timesofindia.indiatimes.com/india/with-12-of-global-imports-india-tops-list-of-arms-buyers-report/ articleshow/63276648.cms; accessed on 26 March 2019.

influence of China in South Asia and beyond. Singh and Rossow outline opportunities for cooperation between the two countries: 'Strengthening maritime domain awareness mechanisms, synergizing ISR assets, enhancing anti-submarine warfare capability, improving the efficacy of our novel cooperative mechanism (the Defense Technology and Trade Initiative), and concluding interoperability agreements (Communications Compatibility and Security Agreement (COMCASA) and Bilateral Exchange and Cooperative Agreement (BECA)'.[3] While it seems that the US would like India to become its ally in a formal sense, the Indian government has opted for a deepening relationship that is nonetheless based on preserving its 'strategic autonomy'. This reflects India's long-standing strategy of pursuing its freedom to exercise maximum options in the context of its dealings and actions with the outside world.[4] Only time will tell the extent to which the Indian government is able to preserve this autonomy and forge an independent position on matters related to both its national and international economic priorities.

While bilateral military and economic cooperation has increased over the last decade, the relationship between India and the USA remains contingent and contested when it comes to the geopolitics of information. It is clear that the Indian IT economy remains dependent on the US market for a large percentage of its export revenues, and that the US IT sector has used the availability and supply of cheap labour from India to its advantage. The various chapters in this book have highlighted how the bilateral relationship on policy matters related to the Internet and IG, IP, and other issues remains fraught. This is especially the case in the matter of IP, where, notwithstanding policies and investments in structures, processes, and policing, the digital has become the basis for multiple productive processes in the informal economy. It is also on the issue of IP that one can clearly see the Indian government's exercise of

[3] Singh, H.K. and R.M. Rossow. 2018. 'Re-shaping the India-US Defence Cooperation in the Indo-Pacific', *The Diplomat*, 24 August, available at https://thediplomat.com/2018/08/re-shaping-india-us-defense-cooperation-in-the-indo-pacific/; accessed on 26 March 2019.

[4] See Upadhyay, S. 2015. 'The Indo-Pacific and the Indo-US Relations: Geopolitics of Cooperation', Institute of Peace and Conflict Studies, available at http://www.ipcs.org/issue-brief/china/the-indo-pacificnbsp-amp-the-indo-us-relations-geopolitics-of-256.html; accessed on 12 August 2018.

its 'strategic autonomy' at the geopolitical level, reflected in the fact that
the Indian government, through its sins of omission and commission, has
been unable to control many manifestations of the digital in India.

A Recalcitrant State?: IP and the Public Interest

India's stance at WIPO and its subsequent changes to the Copyright Act
and the Indian Patents Act have introduced an additional step, namely,
'enhanced efficacy', for patent applicants, which suggests that it is difficult
to assess the country's commitment to IP in black-and-white terms. Large
investments are being made to modernize IP offices in India, upgrade the
facilitation of IP, improve local IP capacities, invest in electronic filing
capabilities, and create policy; and as mentioned earlier, the first draft of a
National IP Policy was released in December 2014.   It is debateable as
to whether the moves to strengthen and harmonize IP laws will result in
the protections that MNC's such as Monsanto, Bayer, and others have
desired in India. IP enforcements are lax illustrated by the fact that it
was during the present PM Narendra Modi's tenure as chief minister of
Gujarat that copies of transgenic seed became widely marketed, shared,
and planted in that state. In fact the GoI has produced at least four
reports that advise against the adoption of GM food crops, although
farmers have still been planting GM crops illegally.[5] The decision taken
by Monsanto, in 2016, to withdraw its new GM cottonseed from India
in protest against the government pressurising Monsanto to share its
proprietary technology with local seed companies does suggest that the
GoI continues to be involved in playing an elaborate IP game that is based
on both openness and closure.[6] Given the pivotal role played by farmers
as a vote base, this ambivalence by the State, reflected in its inability to
enforce IP in agriculture, is entirely to be expected. The lawsuit filed
against the commercial roll out of the first GM food crop in India,

[5] Todhunter, C. 2017. 'GM Food Crops Illegally Growing in India',
Counterpunch, 10 November, available at https://www.counterpunch.org/2017/
11/10/gm-food-crops-illegally-growing-in-india/; accessed on 11 August 2018.

[6] Bhardwaj, M. 2016. 'Exclusive: Monsanto Pulls New GM Cotton Seed
from India in Protest', *Reuters*, 24 August, available at https://www.reuters.com/
article/us-india-monsanto-exclusive/exclusive-monsanto-pulls-new-gm-cotton-
seed-from-india-in-protest-idUSKCN10Z1OX; accessed on 12 August 2018.

transgenic mustard, supported by farmer's unions in India such as the Bharatiya Kisan Union and NGOs such as the Gene Campaign suggests that such issues will remain contentious for the foreseeable future.[7]

While the present government's election manifesto and trade orientation have included pro-IP statements, they will find it difficult to backtrack on the cultures of use related to the digital in the informal sector. The SJM has, on one more than one occasion, objected to US pressure to dilute India's patent laws in favour of global pharma companies as it has against US on India's e-commerce policy.

In 2019, the SJM contested a Supreme Court ruling that granted a patent to Monsanto for its GM cotton seed.[8] This advocacy by the SJM is also visible in other areas. As Bagchi:

> The SJM has ... objected to the policy amendments in 'compulsory licensing' especially while granting permission to NATCO for manufacturing cancer drugs which was being sold by BAYER of Germany 'at exorbitant prices' in the country ... red-flagged 'illegitimate demand of data exclusivity on pharmaceuticals' whereby the Drug Regulatory Authority of India will be prohibited to disclose trial results to the Indian generic companies. SJM has ... flagged a whole lot of other areas where US is seen as trying to mount 'pressure' [and where,] the government should demand 'protection' for products like Darjeeling tea, Basmati rice, textile goods and several other agricultural products which have its origin in India.[9]

---

[7] See Kumar, S. 2017. 'India's First GM Food Crop Held Up by Lawsuit', *The Scientific American*, 18 January, available at https://www.scientificamerican.com/article/india-rsquo-s-first-gm-food-crop-held-up-by-lawsuit/; accessed on 26 March 2019.

[8] See 'Group Close to India's Ruling Party Seeks Change in Patent Act after Pro-Monsanto Court Verdict', 2019. *Reuters*, 8 January, available at https://www.reuters.com/article/us-india-monsanto-patent/group-close-to-indias-ruling-party-seeks-change-in-patent-act-after-pro-monsanto-court-verdict-idUSKCN1P20HY'; accessed on 26 March 2019.

[9] Bagchi, S. 2015. 'US "Threats" to Patent Act Irks RSS Outfit', *The Hindu*, 23 January, available at http://www.thehindu.com/news/national/us-threats-to-patent-act-irks-rss-outfit/article6815593.ece; accessed on 7 August 2018. Also see Datta, J. 2015. 'RSS Economic Wing Seeks Public Debate on Proposed IP Policy', *The Hindu Business Line*, 9 April, available at http://www.thehindubusinessline.com/economy/rss-economic-wing-seeks-public-debate-on-proposed-ip-policy/article7085641.ece; accessed on 7 August 2018.

Incidentally, the farmers' wing of RSS is also against the Land Acquisition Act that would lead to the forcible expropriation of farming land[10] and the introduction of GM crops.[11] Given the reality of RSS membership that includes small farmers, it is not at all surprising that they put forward a variety of positions on IP and agriculture that reflect the needs of its diverse constituencies.

The central government does occasionally take an antagonistic stance against civil society and social movements: for example, proscribing international NGOs such as Greenpeace in the wake of anti-nuclear and anti-development protests and local NGOs that are recipients of foreign aid. Also, as the key initiator of development in the country, the GoI does occasionally stand up for its rights as a sovereign and independent nation. A clear example of such a stance is the commitment of the central government and some state governments, such as in Kerala and Karnataka, to public-sector software based on the principles of FOSS for extending universal access to education.[12] This would suggest that in India, as is also the case in China and Brazil, the State adopts multiple strategies to contest the hegemonic nature of the geopolitics of information.

The following are some examples of the State exercising its 'strategic autonomy' in the matter of IP:

1. In the context of harmonizing its IP legislations with global standards, the State has opted to include clauses that override the interests of IP owners and explicitly support the public need for affordable access to information and inexpensive medical drugs.
2. While it does publicly commit to enforcing IP, it has not, at any given time, been involved in enforcing maximalist versions of IP enforcement as illustrated by the case of copy transgenic seeds.

[10] Yadav, S. 2015. 'There's Pressure from Within Too: Red Flag from RSS Leader', *The Indian Express*, 23 February, available at http://indianexpress.com/article/india/india-others/theres-pressure-from-within-too-red-flag-from-rss-farm-chief/; accessed on 9 August 2018.

[11] 'Sangh Parivar Affiliates Voice Concern over Some Narendra Modi Government Policies', *DNA*, 28 October 2014, available at http://www.dnaindia.com/india/report-sangh-parivar-affiliates-voice-concern-over-some-narendra-modi-government-policies-2029924; accessed on 11 August 2018.

[12] Thomas, P.N. 2014. 'Public Sector Software and the Revolution: Digital Literacy in Communist Kerala', *Media, Culture & Society* 36(2): 258–68.

3. India is also home to a large market in counterfeit goods. A report jointly written by the International Chamber of Commerce (ICC), Business Action to stop Counterfeiting and Piracy (BASCAP), and the Federation of Indian Chambers of Commerce and Industry (FICCI) states that India lost an estimated Rs 72,969 crores in 2012 across seven sectors, namely, auto components, alcohol, computer hardware, FMCG (Personal Goods), FMCG (Packaged Food), mobile phones, and tobacco.[13] The legal scholar, Arul George Scaria, underscores the 'highly cautious approach from the side of Indian judiciary when it comes to standard of proof required for invoking criminal remedies under copyright law.'[14]

4. The GoI is involved in the parallel importation of products such as cell phones from China. These phones are available unbranded and are also sold in India under a variety of brand names, including Spice, Lava, Zen, among numerous others. As Hennessey suggests in an article on the *shanzhai* phones: '*Shanzhai* behavior is not necessarily *against* the law; it is just outside of the government's control' (emphasis in original).[15]

5. The GoI intentionally adopts and adapts FOSS-based platforms in e-governance, state-wise educational initiatives, and invests in both central and state-funded institutions committed to exploring FOSS-based solutions.

In May 2014, the Government of Kerala instructed 'all government and quasi-government institutions migrate their desktop PCs to FOSS-based platforms, unless there were unavoidable reasons to defer the migration. The decision has been taken in view of the withdrawal of support to

---

[13] 'India: Counterfeiting, Piracy and Smuggling in India: Effects and Potential Solutions'. 2013. ICC, BASCAP, FICCI, Paris, p. 8, available at https://cdn.iccwbo.org/content/uploads/sites/3/2016/11/Counterfeiting-piracy-and-smuggling-in-India-Value-of-IP-in-india.pdf; accessed on 26 March 2019.

[14] Scaria, A.G. 2014. *Piracy in the Indian Film Industry: Copyright and Cultural Consonance*. Cambridge: Cambridge University Press, p. 9.

[15] Hennessey, W. 2012. 'Deconstructing *Shanzhai*—China's Copycat Culture: Catch Me If You Can', *Campbell Law Review* 34: 661.

[16] 'Kerala Issues Handbook on Migrating to Free Software Platform', 2014. *Mathrubhumi.com*, 3 June, available at http://www.mathrubhumi.com/english/news/india-elections-2014/kerala-issues-handbook-on-migrating-to-free-software-platforms-147146.html; accessed on 7 August 2018.

Windows XP by Microsoft.'[16] Apart from significant investments in FOSS-based educational initiatives, the government has invested $35 million in the Open Source Drug Discovery (OSDD) project under the aegis of the Council for Scientific and Industrial Research.[17] The aim of this project is to create FOSS-based global initiatives to create drugs to combat tropical diseases, in particular cure for tuberculosis that results in millions of more deaths in India than human immunodeficiency virus (HIV)/AIDS. A proposal from the OSDD to the WHO states: 'The new drug that is likely to come out of the drug discovery process will be made available as a "generic" molecule, free of intellectual property (IP) constraints for the industry to manufacture and distribute anywhere in the world, thereby ensuring that the prices are affordable.'[18]

These examples of contestation suggest that in spite of international pressure, large countries such as India do have opportunities to contest the global geopolitics of information and adapt information for local purposes and in response to local needs. However, the struggle to achieve this is both real and difficult in the face of strong lobbies and bilateral pressures. For example, in the context of the 'free laptop scheme' in the south Indian state of Tamil Nadu, the original plan was to include FOSS software. Close to 2.3 million laptops had already been distributed by early 2014 and another 550,000 units were to be distributed between 2014 and 2017 as part of a $2 billion scheme.[19] A report in the investigative magazine

[17] See Bhardwaj, Anshu, Vinod Scaria, Gajendra Pal Singh Raghava, Andrew Michael Lynn, Nagasuma Chandra, Sulagna Banerjee, Muthukurussi V. Raghunandanan, Vikas Pandey, Bhupesh Taneja, Jyoti Yadav, Debasis Dash, Jaijit Bhattacharya, Amit Misra, Anil Kumar, Srinivasan Ramachandran, Zakir Thomas, Open Source Drug Discovery Consortium, and Samir K. Brahmachari. 2011. 'Open Source Drug Discovery: A New Paradigm of Collaborative Research in Tuberculosis Drug Development', *Tuberculosis* 91: 479–86.

[18] OSDD. n.d. '"Open Source Drug Discovery"': An Open Collaborative Drug Discovery Model for Tuberculosis', Proposal submitted to the WHO Expert Working Group on R&D Financing, p. 1, available at http://www.who.int/phi/public_hearings/second/contributions/Zakir ThomasCouncilofScientificIndustrialResearch.pdf; accessed on 6 August 2018.

[19] Kumar, S.M. 2014. 'Tamil Nadu Budget: 5.5 Lakh Students to Get Free Laptop', *The Times of India*, 13 February, available at http://timesofindia. indiatimes.com/city/chennai/Tamil-Nadu-budget-5-5-lakh-students-to-get-free-laptops/articleshow/30351828.cms; accessed on 7 August 2018.

Tehelka points out that this change from a FOSS-based platform to the proprietary Windows platform was a consequence of a state visit by the then US Secretary of State, Hillary Clinton, to Tamil Nadu, between 10 and 11 July 2011, where she had met the chief minister of the state.[20] Despite this significant turn towards proprietary software, the trend in India, at the level of state governments, is a steady move towards FOSS-based solutions. In Tamil Nadu, in the context of Microsoft not being able to provide technical assistance for the Windows XP operating system and a directive from the IT department that mandates the adoption of FOSS-based solutions by state governments, the government has advised all its departments to install the BOSS Linux platform that has been developed by the C-DAC.[21]

Christopher Kelty has described the role played by 'recursive publics' in the making of FOSS, in particular their capacities to reorient 'power and knowledge'.[22] Revati Prasad, in an article, has highlighted the role played by these 'recursive publics' in the context of Save the Internet campaign organized by the civil society in India against Facebook's Free Basics.[23] We have here an example of a loose community of people who have used their expertise to imagine alternatives in the production of knowledge. However, while acknowledging their efforts at democratizing access to knowledge, we need to take note of the fact that much of these efforts continue to be driven by 'English-speaking' communities located in both the north and the south, who wittingly and otherwise contribute to the

[20] Manish, S. 2011. 'The Deadly Microsoft Embrace', *Tehelka*, 10 October, available at http://tehelka.com/story_main50.asp?filename=Ws101011 MICROSOFT.asp; accessed on 17 June 2012.

[21] Ravi Kumar, N. 2014. 'State Departments Asked to Switch over to Free and Open Source Software', *The Hindu*, 18 March, available at http://www.thehindu.com/todays-paper/tp-national/tp-Tamil Nadu/state-departments-asked-to-switch-over-to-free-open-source-software/article5798429.ece; accessed on 5 August 2018.

[22] Kelty, C.M. 2005. 'Geeks, Social Imaginaries and Recursive Publics', *Cultural Anthropology* 20(2): 185–214; and Kelty, C.M. 2008. *Two Bits: The Cultural Significance of Free Software*. Durham and London: Duke University Press.

[23] Prasad, R. 2017. 'Ascendant India, Digital India: How Net Neutrality Advocates Defeated Facebook's Free Basics', *Media, Culture & Society* 40(3): 415–31.

extension of existing access divides. Gautam John, in an essay on the status of Wikipedia in India in a book that critically interrogates a decade of Wikipedia, has observed that despite the development of Indian language Wikipedias in Hindi, Telugu, Marathi, Bishnupriya Manipuri, Tamil, and Malayalam, and some interest from the government, this initiative remains largely English-centric:

> There have been some technical challenges around the historical lack of growth in Indic language Wikipedia's, in particular in the area of openly licensed and freely available Indic font, difficulties with the cross-platform display of Indic text, and the lack of standardised cross platform Indic language text entry tools. ... This inequitable distribution of content, skewed towards English and languages of the traditional geographies of the global north, has been a frequent point of discussion for the Wikipedia Foundation.[24]

While FOSS, including its practices and environments of use, has provided alternatives to ownership and use, and is providing business solutions in a range of productive sectors across the global economy, the digital, by its very nature, invites multiple illegalities—downloads, copies and its distribution on a global scale from stealth seeds, generic drugs, to the latest Hollywood films and version of Windows. The curbing of what is termed 'cultural piracy' is, of course, a major plank of the foreign policy of Western nations, although the culture of the copy is a lot more complex than what the USTR would like us to believe. In the context of India, for example, the informal market in transgenic seeds highlights the fact that around the edges of capitalism, there are numerous appropriations that are difficult to control because, by their very nature, they are bound to be price sensitive to local needs and as such are bound to be integrated into the local economy and local politics. If the village headman and dominant landowner is using transgenics, then it is not an issue for local farmers, although it certainly will be for Monsanto. This also seems to suggest that though countries in the South are being pressured to harmonize their patent and copyright laws with global standards, the existence of such legislations does not imply its enforcement and compliance. Herring

---

[24] John, G. 2011. 'Wikipedia in India: Past, Present, Future', in G. Lovink and N. Tkacz (eds), *Critical Point of View: A Wikipedia Reader.* Amsterdam: Institute of Network Cultures, pp. 285, 286.

points some of the fallacies that are often manufactured by NGOs of the peasant who has no agency:

> Urban cultural bias resists crediting farmer skill and agency. For example, the rural cottage-industry production and diffusion of dozens of illegal transgenic cotton varieties under the radar of Delhi and Monsanto implies a very different view of the farmer than that of the gullible and hapless peasant. The international oppositional narrative selects for supine peasants and monopolistic multinational corporations with patents.[25]

The rumour of 'terminator' seeds, as much as the activism of farming lobbies and anti-GM activists in India such as Vandana Shiva, plays an important role in keeping conflict levels high. When the government itself is divided among its pro-GM and anti-GM lobbies, it is difficult to enforce legal sanctions against farmers who are mixing and matching seeds, including those that have been digitally altered by Monsanto and in a local laboratory located in some alleyway in an Indian city.

I think one can argue that there are many different approaches to digital gifting that are outside of the dominant pricing system. These certainly include those who contribute to Wikipedia, but also those who freely share transgenic seed. While the latter case is bound up in the discourse and moralities of IP, what it does highlight is the fact that there are communities who believe that there are resources critical to human survival, which simply must not be commodified. The fact that farmers in India are adapting transgenic seed, circulating and sharing it, is a pragmatic response from farmers who are keen to use seed that increases production. However, such acts are also a reminder to us that in the context of highly stratified and unequal contexts of global development, sufficient numbers of people accessing informational products, adapting these products, and using it in the context of local survival strategies need to be seen as correctives to an uncontrollable capitalism.

The polymorphous State in India is, in its present incarnation, committed to capitalism and capitalist development. The present government's development objectives include the transformation of India into a modern country in which everyone, including the poor, becomes

---

[25] Herring, R. 2009. 'Persistent Narratives: Why Is the "Failure of Bt Cotton in India" Story Still with Us?', *AgBioForum* 12(1): 19.

digitally literate and digitally wise. How to achieve this in the context of other divides in the society is, of course, a major blind spot given that there is a belief that digital access will magically resolve these other divides. The dominant approach is based on welcoming digital capitalism to India, mainstreaming digital productivity across all productive sectors, and transforming India into a hub for a host of back-end operations by using cheap labour costs to its advantage. While the available narrative circulated for global consumption focuses on the success story of software entrepreneurs, programmers, and software exports and exporters, the development of the knowledge economy in India is also a story of land acquisitions by software companies, often with the help of the State, for throwaway prices, the continuing struggles for compensation for those who lost their lands, and poor labour conditions, especially for workers in the BPO sectors. The SEZs, IT corridors, and factories have been built on both the commons and farmlands. The knowledge economy is being built through the displacement of traditional agricultural economies via legislations, such as the Land Acquisition Act.

The 'Land' Issue: Politics and Geopolitics

Since geopolitics is also about how global policy impacts the local, I would like to conclude with some observations on how one of the key factors related to capitalism in India, that is, the ceaseless quest for land and its resources—or ABD—is affecting one of the most vulnerable communities in India, the Adivasi's (indigenous people). Let me begin by affirming that India's embrace of the knowledge economy, the growing contribution by this sector to the GDP, the penetration of mobile phones, investments in smart cities, along with all other aspects of the story of information growth, is fundamentally the manifestation of uneven growth. All this continues to impact a small section of the Indian population, although the cumulative impact of the current land grab by the State, IT, mining, and real estate has begun to impact large sections of its population. The discourse of growth is bounded by neo-liberal economics and, therefore, there is little emphasis on a larger understanding of growth based on distributive justice, and equality, that strengthens the life chances of India's most vulnerable communities. Arguably, the Indian public has been introduced to an attenuated understanding of growth through means such as the Land Acquisition Bill. In its previous avatar as the Land

Acquisition, Rehabilitation and Resettlement Act, 2013, both the public and private sectors had to obtain consent from the majority of farmers for land acquired for private and public–private investments. The present government has tried to do away with the prior need for consent, or for social impact assessments, for land to be used for security projects and infrastructure projects such as roads and dams, and SEZs.

Resistance to such measures have, typically, been dismissed as a law and order problem—and yet it is precisely the most poverty-stricken and insurgent regions that have been slated for land grabs by both the public and private sectors. In Chhattisgarh, a state in central India that is the epicentre of the Maoist movement, 20,000 acres of land in just one district—Raigarh—has been allocated to mining.[26] Adivasis in India have been relentlessly marginalized during the post-Independence era and their lands expropriated by the State by its reason of 'eminent domain'. Close to 80 per cent of Adivasis live in eastern and central India on land that is rich in minerals. A number of official and independent investigations have revealed the ruin of local economies and ways of living as a consequence of the expropriation of tribal lands by the State and private sector. A report by the Centre for Environment and Food Security presents these consequences in the starkest of terms:

> Out of a total 1000 sample Adivasi households from 40 sample villages in Rajasthan and Jharkhand surveyed for this study, a staggering 99 per cent were facing chronic hunger. The data gathered during this survey suggests that 25.2 percent of surveyed Adivasi households had faced semi-starvation during the previous week of the survey. This survey found that 24.1 percent of the surveyed Adivasi households had lived in semi-starvation conditions throughout the previous month of the survey. Over 99 per cent of the Adivasi households had lived with one or another level of endemic hunger and food insecurity during the whole previous year. Moreover, out of 500 sample Adivasi households surveyed in Rajasthan, not a single one had secured two square meals for the whole previous year.[27]

[26] Podur, J. 2013. 'The Bastar Land Grab: The Expropriation of Farmers in India', *Global Research*, 22 April, available at http://www.globalresearch.ca/the-bastar-land-grab-the-expropriation-of-farmers-in-india/5332410; accessed on 22 August 2018.

[27] *Political Economy of Hunger in Adivasi Areas*. 2005. Report prepared by the Centre for Environment and Food Security, New Delhi, available at http://

Their lack of access to health facilities is highlighted by Baru et al.;[28] and Guha, in a study on Adivasis, Naxalites, and Indian democracy, has highlighted the systematic ways in which tribal lands have been alienated, their low literacy rates, indebtedness, and lack of access to health and drinking water.[29] Yogesh Jain, writing from the heart of conflict in Chhattisgarh, states that 'the largest "health problem" that we see is of rampant under nutrition, a polite term for chronic hunger. More than 70 per cent of children below the age of three, and more than half of all adults in this part of rural Chhattisgarh are undernourished. Shockingly, these numbers don't elicit much outrage.'[30]

The most telling indictment of the system, however, is contained in the official report—*Committee on State Agrarian Relations and Unfinished Task of Land Reforms, Vol. 1: Draft Report*.[31] The report includes information on the variety of ways in which tribal lands have been alienated but reserves its strongest critique for the involvement of the private sector. The concluding section of Chapter 4 in their report titled 'The Biggest Land Grab of Tribal Lands after Columbus', states the following:

A civil war like situation has gripped the southern districts of Bastar, Dantewada and Bijapur in Chhattisgarh. The contestants are the armed squads of tribal men and women of the erstwhile Peoples War Group known as the Communist Party of India (Maoist) on the one side and the armed tribal fighters of the Salwa Judum created and encouraged by the government and supported with the firepower and organisation of the central reserve police forces. The open declared war will go down as the biggest land grab ever, if it plays out as per the script. The drama being scripted by Tata Steel

---

www.cefs-india.org/reports/Research%20Study%20on%20the%20Political%20 Economy%20of%20Hunger%20in%20Adivasi%20(Tribal)%20Areas%20of%20 India.pdf; accessed on 7 August 2018.

[28] Baru, R., A.B. Acharya, S. Acharya, A.K.S. Kumar, and K. Nagaraj. 2010. 'Inequities in Access to Health Services in India: Caste, Class and Region', *Economic and Political Weekly* XLV(38): 49–58.

[29] Guha, R. 2000. *The Unquiet Woods: Ecological Change and Peasant Resistance in the Himalaya*. Berkeley: University of California Press.

[30] Jain, Y. 2012. 'Healthcare in an Uneven World', *Tehelka*, 30 June, p. 33.

[31] *Committee on State Agrarian Relations and Unfinished Task of Land Reforms*. 2008. *Vol. 1: Draft Report*. New Delhi: Ministry of Rural Development, GoI.

and Essar Steel who wanted 7 villages or thereabouts, each to mine the richest lode of iron ore available in India. ... Villages sitting on tons of iron ore are effectively de-peopled and available for the highest bidder.[32]

Just as it is important to deal with the role of the State in the shaping of India's knowledge economy and its role in negotiating the geopolitics of information, as I have tried to deal with in this study, it is also important that we do not forget that this story is also about the everyday lives of people who have been either affected directly by India's IT economy, such as low-paid women workers in India's call centre (BPO) industry,[33] as much as the Adivasis and other rural workers whose farms and lands have been expropriated in order to build this new and smart India.

Neo-liberal development in India has been based on a deep marketization and the increasing role played by financialization. Harvey has described financialization as the growing influence and expansion of a range of financial institutions, actors, instruments, and devices within a deregulated financial environment.[34] In the context of India, the financing of IT corridors and smart cities is not only dependent on public funds but also private equity and, to a lesser extent, on FDI and money raised through multilateral and bilateral means. In the case of the capital city of AP, Amaravati, the total cost is in the region of Rs 42,935 crores (US$9.2 billion) that will be raised through the sale of public bonds, the involvement of firms such as McKinsey to raise and mobilize private and global finance, and effected through PPPs, annuity, and BOT models. Given the increasingly important role played by finance companies that often belong to large conglomerates that are also involved in real estate—for example, L&T Finances that is part of Larsen & Toubro Limited, one of the leading companies in India with interests in engineering, construction, electrical and electronics manufacturing and services, IT, and financial services—financialization clearly is a corporate strategy aimed at the maximization of flexible accumulation.

---

[32] Committee on State Agrarian Relations and Unfinished Task of Land Reforms, pp. 160–1. See also Laul, R., 'How to End a Million Mutinies', Tehelka, 13 August 2011, pp. 40–5.

[33] See Patel, R. 2010. Working the Night Shift: Women in India's Call Centre Industry. Stanford: Stanford University Press.

[34] Harvey, D. 2006. Spaces of Global Capitalism: Towards a Theory of Uneven Geographical Development. London & NY: Verso, pp. 24–5.

The example of Satyam Computers after its attempts to diversify into real estate, through the family-owned Maytas Properties that owned 6,800 acres of land in south India, is suggestive of the extent of money laundering that is transacted via 'shell' companies, including those that are 'used for rotation and siphoning off of funds', 'used for creation of equity in their name', and 'used for holding real estate properties'.[35] In the case of Satyam Computers, the chairman Ramalinga Raju's taste for real estate contributed to the fall of a once iconic software company. As Venkateshwarlu observed: 'It remains a mystery how Raju managed to acquire 6,800 acres of land, with preliminary reports suggesting benami (assumed names) purchases or the floating of a number of companies to circumvent laws. On paper, it does not go beyond 1,000 acres, including IT special economic zones and fancy villa and apartment projects around Hyderabad.'[36]

While there have been billions of dollars worth of investments in the Digital India initiative that was launched in 2015, there is growing evidence that its ambitious projections have run up against the reality of a country characterized by many divides between the rich and the poor and uneven infrastructures across regions as well as social groups. It has become clear that investments in digital literacy cannot stand alone and must be connected with investments in infrastructure accessible to all, irrespective of caste and creed. Digital India and its core drive towards a cashless India had been infamously imposed on people by demonetization in 2016. As with other aspects of the Digital India programme, this policy had a differential impact on populations across India, with the poor bearing the brunt of disruptions in their everyday life given their dependence on hard cash for their everyday survival.[37] When it comes to other aspects of Digital India, the National Digital Literacy Mission

[35] Singh, D. 2010. 'Incorporating with Fraudulent Intention', *Journal of Financial Crime* 17(4): 472.

[36] See Venkateshwarlu, K. 2009. 'Maytas Twins', *Frontline* 26(3), available at https://frontline.thehindu.com/static/html/fl2603/stories/2009 0213260301000.htm; accessed on 26 March 2019.

[37] Sidhu, J. 2017. 'Alongside Modi's Digital India, a Mounting Pile of Unanswered Network Quality Complaints', *The Wire*, 26 November, available at https://thewire.in/196971/alongside-modis-digital-india-mounting-pile-unanswered-network-quality-complaints/; accessed on 9 August 2018.

(2014–16) and its successors, Digital Saksharta Abhiyan (DISHA 2016) and the Pradhan Mantri Gramin Digital Saksharta Abhiyan (2017), have achieved mixed results. Evaluation studies have revealed the duplication of beneficiaries, significant shortfalls in the training of people belonging to the Scheduled Castes and Scheduled Tribes, and more importantly, the fact that to succeed, digital literacy is linked to people's access to the Internet and relevant devices.[38]

The incongruence of the vision of Digital India with the lives of the poor in India is perhaps best illustrated by a report from Kalahandi, Odisha, where the 'Hello Point' in the village is the only place where one can get phone signals. Sanjiv Phansalkar thereby places the Digital India initiative in perspective and his words are a stark reminder that the political economy of uneven development simply has to be factored into an understanding of India's embrace of the digital:

> So when one goes … with someone having a smartphone to the Hello Point and when the mobile data network is functioning, then the people can perhaps enjoy the formidable privileges of digital inclusion. Literacy is commensurate with the level of poverty here. Surely very few people know how to read and write in English. How would they transact digitally with sites which are largely in English is any one's guess.[39]

***

In this book, I have tried to highlight the relationship between the global and the local in the structuring of the geopolitics of information. While this concluding chapter has highlighted the Indian state's 'strategic autonomy' on matters related to IP and FOSS, these positions

---

[38] Jain, M. 2017. 'India's Ambitious IT Literacy Plan is Stumbling over Poor Infrastructure and Faulty Processes', *Quartz India*, 30 October, available at https://qz.com/1114895/digital-india-indias-ambitious-it-literacy-plan-is-stumbling-over-poor-infrastructure-and-faulty-processes/; accessed on 20 August 2018.

[39] Phansalkar, S. 2018. 'India's Rural Poor Haven't Benefited Even a Little from the Famed Digital "Revolution"', *Huffington Post*, 1 January, available at https://www.huffingtonpost.in/village-square/indias-rural-poor-havent-benefitted-even-a-little-from-the-famed-digital-revolution_a_23322489/; accessed on 10 August 2018.

in themselves reflect the complex relationship that it has with a variety of global and local issues related to the geopolitics of information. The present State, as a keen supporter of neo-liberalism, will undoubtedly extend the market in and through a variety of bilateral and multilateral means. However, as the experience with Free Basics reveals, despite being a major corporate–State initiative, the regulator took sides with popular opinion. India's economic relationship with the USA is multifaceted and complex, and the impact of the increasing turn towards protectionism in Washington on India's IT and IT-enabled service economies is yet to become fully apparent. A lot will depend on whether or not a lot more investments will be targeted towards the strengthening of the domestic knowledge economy, along with the development of indigenous strengths in sectors linked to the knowledge economy. Such investments will be key to India maintaining its competitive advantage. Often described as a democratic bulwark, India is seen as a natural ally of the West and both the State and its middle classes are firm supporters of neo-liberalism. However, as I have tried to highlight in this book, neither the State's agenda nor that of the middle classes necessarily corresponds with the agendas of people working in the informal economy and/or India's many millions of poor. These cleavages and rifts in India's democracy, economy, and society are expressed in multiple ways and encompass a variety of attitudes towards the digital as resource, commodity, and property.

Polanyi's engagement with the 'double movement' was based on his critique of a market economy that had become all-powerful and that had subsumed and subjugated land and labour. While the State in Europe attempted to counter the market through supporting different versions of the welfare economy, in the case of India and the knowledge economy, there are numerous actors involved in countering the excesses of informational capitalism, including the State itself, along with civil society and its many 'jugaad' cultures. Despite its clear embrace of the market economy, the Indian state simply cannot ignore the millions of poor people who depend on the State for their survival and who use their franchise to periodically vote for those who promise to support their economic interests. Governments that err towards the market, as is the case with the present government, are thus vulnerable to this corrective impulse, just as the 'broad class of labour did not shrink from breaking its (the market's) rules and challenging it outright.'[40] The potential for disruptive anti-market initiatives is high in a country that writer V.S. Naipaul once described as a land of a million mutinies.[41] Thus, arguably, India's double movements around the digital are reinforcing the primacy of social, community, and relationships based on reciprocity and mutual exchange—thereby offering a much-needed challenge to ABD.

The agency of the State itself can nonetheless be overstated in this regard. I have argued in this book that the writ of the State is typically only loosely enforced, with the consequence that there are large spaces within both formal and informal economic environments, where there are opportunities for multiple disruptive valuations of information. The contested nature of 'information' needs to, however, be seen in context—the slow transitioning of India from a minor state to an ambitious global player. This transitioning is highly contested and has, in its wake, disrupted millions of people from their lands and livelihoods. Despite the many conflicts in Indian society that are a consequence of caste, class, religion, and gender, and the many seemingly intractable cleavages, its strengths include the fact that Indian democracy, despite claims to its avowed wholeness, is incomplete, imperfect, and fractious. It is this reality that has given rise to multiple social movements, from the Maoists to groups such as the People's Union for Civil Liberties (PUCL), whose aims are to create the conditions for justice, equity, and to introduce a politics that is conducive to a people-based, rather than market-based, strategy of growth. It is clear that these struggles in India are primarily of a local–specific nature, reflecting the fact that much of the contestation against ABD is, in itself, 'diffuse, very much a function of the Inchoate, fragmentary, and contingent forms taken by accumulation by dispossession.'[42] In this sense, there will be numerous challenges to informational capitalism's predations in India that will take many forms, including resistance to the expropriation of land, the fight for better wages and better working conditions for workers in India's information industries, 'jugaad' cultures, and mobilizations for access and affordable connectivities.

[40] Polanyi, K. 1944. *The Great Transformation: The Political and Economic Origins of Our Time*. New York: Farrar & Reinhardt, p. 191.

[41] Naipaul, V.S. 2010. *India: A Million Mutinies Now*. London: Pan Macmillan.

[42] Harvey, D. 2003. *The New Imperialism*. Oxford and New York: Oxford University Press, pp. 173–4.

Index

About the Author

Pradip Ninan Thomas teaches at the School of Communication and Arts, the University of Queensland, Brisbane, Australia. He has written extensively on the media in India, the political economy of communications, communications and social change, and media and religion. Thomas was the vice-president of the International Association for Media and Communication Research from 2012 to 2016. His most recent publication is *Empire and Post-Empire Telecommunications in India: A History* (2019, OUP).